The Applicability of Mathematics as a Philosophical Problem

The Applicability of Mathematics as a Philosophical Problem

Mark Steiner

Harvard University Press
Cambridge, Massachusetts
London, England
1998

*To the memory of my father
and my mother*

Copyright © 1998 by the President and Fellows of Harvard College
All rights reserved
Printed in the United States of America

Library of Congress Cataloging-in-Publication Data
Steiner, Mark.
The applicability of mathematics as a philosophical problem / Mark
Steiner.
p. cm.
Includes bibliographical references and index.
ISBN 0-674-04097-X (alk. paper)
I. Mathematical physics. 2. Mathematics—Philosophy. I. Title.
QC20.S743 1998
530. 15—dc21
98–3468

Contents

Acknowledgments

I am grateful to Daniel Amit, Paul Benacerraf, Bernard Berofsky, Jed Buchwald, Sylvain Cappell, Alan Chalmers, Percy Deift, Burton Dreben, Shmuel Elitzur, Juliet Floyd, Harry Furstenberg, Joel Gersten, Shelly Goldstein, Shaughan Lavine, Jesper Lützen, David Malament, Nathan Malkin, Benoit Mandelbrot, Sidney Morgenbesser, Yuval Ne'eman, Robert Nozick, Itamar Pitowsky, Carl Posy, Hilary Putnam, Sam Schweber, Stewart Shapiro, Barry Simon, Shlomo Sternberg, and Larry Zalcman for their much appreciated advice and criticism. Tragically, I can no longer thank Yael Cohen for all the help she gave me in working on this book. Nessa Olshansky-Ashtar made several editorial suggestions; thanks, Nessa. Angela Blackburn's professional copyediting improved the book markedly; it was a treat to work with her.

This book has had a very long gestation period. In some sense, it is my final attempt to persuade my advisor and mentor, Professor Paul Benacerraf, that there really *is* a philosophical problem about the applicability of mathematics in the natural sciences. His influence on this book consists as much in what I left out as in what I put in.

To many of the people on this list, particularly the physicists and mathematicians, I am indebted for much more than advice and criticism. I could never have written a book on modern physics without them—in some cases, they simply taught me the material. In other cases, they corrected dangerous misconceptions and shared their insights with me. I do not know whether to thank them or to apologize for taking up so much of their valuable time; in some cases, their e-mail correspondence with me was more voluminous than this book. (Which reminds me to thank whoever invented e-mail, without which this book certainly could not have been written.)

The reader will note that some of the most beautiful examples of mathematical reasoning in physics in this book are borrowed from the writings of Professor Shlomo Sternberg. I am grateful to him for the patient way he explained these writings to me, even when it took him four or five attempts till I "got it."

My research was supported by grants from the Israel Academy of Sciences and the United States National Science Foundation; for this support I express my appreciation.

An invitation to the Dibner Institute at the Massachusetts Institute of Technology, for which I am grateful, afforded me the opportunity to consult with eminent scholars there and in the Boston area generally.

Finally, my thanks go to the editors of the *Journal of Philosophy* and *Philosophia Mathematica* for permission to reproduce material from Steiner 1995 and Steiner 1989 in Chapters 1 and 3, respectively.

And to my wife, Rachel, and the rest of the family, I offer thanks for the love and patience which waiting for the book required.

Introduction

This book has two separate objectives.

The first is to examine in what ways mathematics can be said to be applicable in the natural sciences or, if you prefer, to the empirical world. Mathematics is applicable in many senses, and this ambiguity has bred confusion and error—even among "analytic" philosophers: because there are many senses of "application" and "applicability," there are many questions about the application of mathematics that ought to be, but have not been, distinguished by philosophers. As a result, we do not always know what problem they are dealing with.

For example, we often use pure mathematics in reasoning about the empirical world. This raises two questions: what is the *logical form* of such reasoning, and what are its *metaphysical presuppositions*?

The problem of logical form arises because number words function in mathematics as proper nouns (names of numbers, numerals), while in empirical descriptions ("three green leaves") they often function as adjectives.[1] This is an equivocation, which appears to make it impossible to reason using mathematics in an empirical situation. Thus, some writers have felt themselves forced to distinguish between "pure mathematics" and "empirical" or "applied" mathematics. I will point out that Frege solved the problem, making this distinction entirely superfluous, by showing how number words can function as names, not adjectives, everywhere.

Names of what? Names, apparently, of numbers, thought of as abstract or nonphysical objects. And this brings us to the *metaphys-*

---

[1] This is true in English and other European languages, but not in the Semitic languages. Thanks to my brother, Richard Steiner, for clarifying the situation for me.

ical question concerning application: how can facts about numbers be relevant to the empirical world? Frege had a keen answer to this, too: they aren't. Numbers are related, not to empirical objects, but to empirical concepts! It is the empirical concepts that are used to describe the world; numbers are used to characterize those very descriptions. (Of course, there are other objections to Platonism than its alleged inability to account for applications, but this is not a book about Platonism.)

Thus, two frequently asked questions about the application of mathematics were answered definitively over a hundred years ago. If philosophers sometimes write as if this were not so, the reason may be because, as I suggested above, the different concepts of applicability have not been made clear. To do this is precisely my first objective.

* * *

To the extent they have discussed mathematical applicability at all, contemporary philosophers have usually explored its implications for mathematics itself. Following the great physicists, however, I would like to explore its implications for our view of the universe and the place in it of the human mind (or minds like the human mind, if there are any), and this is the second objective of the book.

Galileo, of course, remarked in the seventeenth century that the Universe is a book written in the language of mathematics (actually: geometry). But the best known contemporary essay on the application of mathematics and its meaning for humans is by the celebrated physicist, Eugene Wigner, who spoke of the "unreasonable effectiveness of mathematics in natural science," which is a gift we "neither understand nor deserve." In fact, my main argument concerning the applicability of mathematics, though different from Wigner's, could easily be confused with his. Let me, therefore, outline the main argument of my book and show how it differs from, and escapes common objections to, Wigner's.[2]

[2] I do this at the suggestion of Burton Dreben, Juliet Floyd, and Robert Nozick, whom I had the good fortune to consult with while it was still possible to rewrite the Introduction. Further valuable advice came from Hilary Putnam, Sam Schweber, Alan Chalmers, Jed Buchwald, and Jesper Lützen. The opportunity to meet all these people was made possible by an invitation to spend the year at the Dibner Institute of

At the end of the nineteenth century, physics was at a crossroads. Scientists were attempting to describe the unseen world of the very small, and they already knew that the atomic world obeys different laws than those governing the macroscopic world. Of course, as Charles Peirce pointed out, once the laws were conjectured (abduction), one could verify them indirectly by checking their consequences for the macroscopic world (induction). But first they had to be guessed, and Peirce himself (like John Locke before him) was pessimistic about the ability of the human species even to conjecture the laws of the atom. Evolution, he argued, could not have equipped the human species with the ability to come up with the laws of objects which play no role in our daily life. (This was exactly Locke's argument, except that he said "God" instead of "Evolution.")

How, then, did scientists arrive at the atomic and subatomic laws of nature? My answer: by mathematical analogy. Of course, not only by mathematical analogy and not only the atomic problem. The usual procedure of experimental inquiry went on just as in the past. What was new in the conduct of research was an increased reliance by scientists *also* on mathematical analogies. To be sure, classic inquirers like Newton and Maxwell had already used mathematical analogies, as I will argue. Now, however, scientists used mathematical analogies because they had no real alternative. Scientists looked for laws bearing a similar (not necessarily identical) mathematical form to the laws they were trying to augment, refine, or even replace. Often these analogies were *Pythagorean*, meaning that the analogies were *then* inexpressible in any other language but that of pure mathematics. The Newtonian doctrine that Nature is "conformable to herself" gained a new twist, indeed a considerable expansion of meaning, when the conformability was defined in terms of mathematical analogies.

Even where the analogies assumed the form of apparently physical models (as, for example, the Bohr model of the hydrogen atom as a "planetary" system), I will try to show that these models also functioned as mathematical metaphors. That is, the mathematical form of the models was abstracted out, and then applied analogically even in

MIT. At the risk of repeating the Acknowledgments, I would not like to overlook the roles of Sidney Morgenbesser, Sylvain Cappell, Carl Posy, and other friends.

areas where the actual behavior of the atom could not be described by classical mechanics. As Dirac would later put it, the deep meaning of the so–called "Correspondence Principle" of Bohr was that the quantum laws were to have the same mathematical form as the classical laws.

Thus, any analogy among structures the mathematician did, or could, recognize, became a potentially physical analogy, too. Mathematics itself thus provided the framework for guessing the laws of the atomic world, by providing its own classificatory schemes.

Often, once an atomic or subatomic law was guessed successfully, the analogy at the base of the guess was discovered to have physical warrant. We will see how the theory of quarks, for example, underwrites the analogy used in predicting the omega minus particle by Gell-Mann (Ne'eman also predicted the particle, but I will contend that his reasoning relied less on analogies). This analogy was between an abstract symmetry group already employed in physics, called "SU(2)," and another one called "SU(3)." These two symmetry groups (as the very notation signals) have an obvious mathematical analogy, but at the time that Gell-Mann and Ne'eman guessed the existence of the omega-minus particle, there were actually positive reasons to think that quarks did not exist.

Again, I do not claim that the Pythagorean analogy between SU(2) and SU(3) could have by itself generated the discovery. An enormous amount of data intervened—maybe too much data. But that data was brought to bear on the problem by means of the analogy. In fact, we will see later that Gell-Mann, who was not aware of the relevant mathematics literature, tried to discover SU(3) by himself, by generalizing the mathematical attributes of SU(2). In the end, he had to consult mathematicians who told him that the work had been done long ago.

In some remarkable instances, mathematical *notation* (rather than structures) provided the analogies used in physical discovery. This is particularly clear in cases where the notation was being used without any available interpretation. So the analogy was to the *form* of an equation, not to its mathematical meaning. This is a special case of Pythagorean analogies which I will call "formalist" analogies.

The strategy physicists pursued, then, to guess at the laws of nature, was a Pythagorean strategy: they used the relations between the structures and even the notations of mathematics to frame analogies and guess according to those analogies. The strategy succeeded. *This does not mean that every guess, or even a large percentage of the guesses, was correct*—that never happens on any framework for guessing. What succeeded was the global strategy.

Because my work focuses on "Pythagorean" reasoning, I might inadvertently create the impression that I think that vacuous mathematical manipulation, rather than empirical inquiry, was what formed modern science. So let me say immediately that no scientist mentioned here could have formulated valuable theories without an enormous fund of empirical information and prior modeling. And no discovered theory can be confirmed except by empirical testing, by what are called experiments. My point is that this empirical information was brought to bear on new cases through the medium of mathematical classification. Which is just to make a Galilean point: in formulating conjectures, the working physicist is gripped by the conviction (explicit or implicit) that the ultimate language of the universe is that of the mathematician. (I do not claim, of course, that Galileo shared any of the further views about mathematics which I am about to develop.)

The reader who is unwilling to follow me further can now adopt some version of Pythagoreanism. This is not my position, as you will see, but I consider it to be very respectable. The success of the Pythagorean strategy might lead the reader to *conceptual* Pythagoreanism, the view that the ultimate properties or "real essences" of things are none other than the mathematical structures and their relations. More radically, one might adopt *metaphysical* Pythagoreanism, which simply identifies the Universe or the things in it with mathematical objects or structures. (Some physicists write as though an elementary particle just "is" an irreducible group representation, or even that the entire universe is.)

But I will not rest at Pythagoreanism of any kind. For the main argument of this book is that, given the nature of contemporary mathematics, a Pythagorean strategy cannot avoid being an *anthropocentric* strategy. An anthropocentric strategy for making a discovery is one which makes no sense unless the strategist believes, if only implicitly or unconsciously, that the human species has a special place in the scheme of things. The difference between male and female is important to us as humans, but anyone who guesses that subatomic particles should be classified as male or female—even for purposes of discovery—is *almost certainly* adopting an anthropocentric strategy, one which makes no sense, if the universe is indifferent to our goals and values. (Any appearance of a "caring" universe—as evidenced by the fitness of creatures to their environment—is to be explained away by natural selection, according to Darwinists, and is therefore an illusion. For this reason, the male-female distinction *in biology* does not count as anthropocentric, if based, say, on an assumption that sexual differentiation has survival value to complex forms of life.)[3]

In fact, in the Middle Ages, physics *was,* arguably, anthropocentric—at least covertly—because it classified events into heavenly and terrestrial, a distinction reflecting our own parochial point of view. It is notorious that Freud regarded the geocentric universe as highly anthropocentric, indeed narcissistic. (In Chapter 3, I will discuss whether he was right.)

Now I claim that to use mathematics to define similarity and analogy in physics is almost as anthropocentric as using "male-female" or "earthy-heavenly" as classifying tools. Why? Because the concept of mathematics itself is species-specific. (This is a position which would not have occurred to the Pythagoreans of old, because mathematics then consisted of arithmetic and geometry, and the argument fails if that is all there is to mathematics.) There is no objective criterion for a structure to be mathematics—and not every structure counts as a mathematical structure. Chess, for example, has a structure. But mathematicians do not regard theorems about this specific structure as worth bothering about. (This remark removes from consideration a distressingly common "explanation" for the effectiveness of mathematics in physics: mathematics studies "structures," and these structures are displayed in nature where they can be studied by physics. I believe that the currency of this explanation stems from a confusion among the various senses of "applicability," which I also want to clarify in this book.)

Where mathematicians used to look to utility in science (after all,

[3] I am grateful to Sylvain Cappell for raising the issue of the male–female distinction in biology.

many of them were also physicists), mathematicians *today*[4] have adopted internal criteria to decide whether to study a structure as mathematical. Two of these are *beauty* and *convenience*.

The *beauty* of the theory of a structure is a powerful reason to call it mathematical. Yet what we call beautiful (I argue) is species-specific. Thus, a mathematician who uses beauty as a criterion for mathematicality but then claims that mathematics is universal, suffers from a discrepancy between doctrine and behavior; when studying scientists, I always look to the behavior, not the doctrine.

Moreover, the computing power of our brain is limited, Therefore, calculational *convenience* becomes a reason for studying a concept. Mathematicians, that is, introduce concepts into mathematics to make calculations easier or convenient. If we had brains a thousand times more powerful, we would presumably not need these particular concepts. Thus, both beauty and convenience (as well as related notions like comprehensibility) are anthropocentric, or species-specific, notions.

I am not claiming, for example, that the concept of a *group* is anthropocentric—on the contrary. What is anthropocentric is the concept of *mathematics*. My major claim is that relying on mathematics in guessing the laws of nature is relying on human standards of beauty and convenience. So this is an anthropocentric policy; nevertheless, physicists pursued it with great success.

I noted before that the notation of mathematics, and not just what the notation expresses, also played a role in scientific discovery. Here the anthropocentrism is most blatant. For example, some of the analogies physicists drew were formal, i.e., syntactical: the equations they guessed simply looked like the equations they already had. In such cases, scientists were studying their own representational systems—i.e., themselves—more than nature. Scientific research sometimes looked like eighteenth-century Eulerian symbol manipulation, as when Dirac factored an unfactorable quadratic polynomial to get an equation which was formally linear. I will describe below how Dirac succeeded in interpreting this formal equation as a linear 4×4

---

[4] Amusingly, even when the contemporary mathematician is also a physicist, the former ignores the needs of the latter. For a remarkable statement of this, see Dyson 1972.

matrix equation in (formal) analogy to what Pauli had done before him with 2×2 matrices.

My claim is that an anthropocentric policy was a necessary factor (*not* the only factor) in discovering today's fundamental physics. This makes the universe look, intellectually, "user friendly"[5] (in that our categories of beauty and convenience are found in the "real essences" of things) to our species, or other species like ours, if any. I say: to our species, rather than to the community of scientists, because the kinds of moves I will document in this book are the kinds of reasoning (anthropocentric, magical) which are common to all children. This does not mean that children could have discovered the Dirac equation; it means that the ability to see things as a child would is one which the great scientists have not lost in all their sophistication.

I mean this claim, that ours is (or rather: appears to be) an intellectually "user friendly" universe, a universe which allows our species to discover things about it—I mean this claim to stand as an empirical hypothesis, and as the conclusion of this book. But though the conclusion is empirical, much philosophy is needed to cut through the difficult issues to see that there *is* an empirical phenomenon here in the face of attempts to explain away the data.[6] Let me discuss briefly a few such attempts.

To explain my data away, one must find a natural, or material, property of mathematics as such, and then show how this property accounts for the success of the mathematical discoveries to be outlined below. For example, a Darwinist account of the origins of mathematics in prehistory (of which there are now several) is not enough to refute the main argument of this book; on the contrary, it might end up confirming it. To show this most clearly, I will consider the most egregious forms of scientific guessing in this century, the "formalist" moves that relied on the appearance of the mathematical notation (i.e., its "syntax"). Recall the symbol manipulation of Dirac, mentioned above. The preference for patterns in nature might well have been selected for, and it is obvious how such a preference

[5] Thanks to George Schlesinger for this phrase.

[6] The distinction between explaining a hypothesis and explaining it away is originally due to my mentor, Sidney Morgenbesser. In explaining away, one explains the appearance of a phenomenon, rather than its existence.

might spill over into a preference for the same patterns in mathematical notation. This explains something about mathematical notation, but not why it is successful. On the contrary, a similar explanation could be given for the use of palindromes in ancient magic—palindromes are sentences spelled the same way forward and backward, and we prefer symmetrical objects. This explanation explains the existence of magic, but also explains why we would not expect it to work.

Some philosophers prefer a Kantian account of mathematical discovery: the world is the way it is, in part because of our contribution to our own experiences. Mathematics is the lens through which we view the Universe, meaning the phenomenal, or experienced, Universe (about things in themselves we know nothing). This is also a valid attempt to explain away the data, but it will have to come to grips with the nature of contemporary science, which deals with objects beyond the realm of spatiotemporal experience.

Let us now look at the difference between my project and that of Wigner.

Wigner speaks not of discovery, but of description: he asks, why is it that the concepts of mathematics (of all things) pop up in physical laws? There are two problems with his "mystery." First, he ignores the failures, i.e., the instances in which scientists fail to find appropriate mathematical descriptions of natural phenomena (which outnumber the successes by far). He also ignores the mathematical concepts that never have found an application. A deeper problem with Wigner's formulation is: each success of applying a mathematical idea to physics is just that—an individual success of a mathematical concept. The success of the group concept, for example, might have nothing to do with a group being a *mathematical* concept. My own formulation avoids both problems: what has been astonishingly successful was a grand strategy, not an isolated act, and what succeeded was the use of the entire structure of mathematical concepts, not this or that concept. Since this structure is defined anthropocentrically, we can now conclude both that physicists acted as though they held (implicitly, for the most part) anthropocentric beliefs, and also that the world really does look anthropocentric—in the limited sense that it is intellectually accessible to human research.

To the extent that "naturalism" rejects any anthropocentric point of view (and I think all forms of naturalism do)[7]—then this book challenges naturalism. This makes the book consistent with natural theology, but there are many positions available that are neither naturalist nor theological. I trust that readers who have no interest in, or commitment to, theology will be able to learn things about the application of mathematics in the natural sciences from this book. I know one reader who was converted by this book, not to theism, but—despite my best efforts—to Pythagoreanism.

In my research I have been inspired in great measure by Maimonides' view that no philosophy, and in particular no religious philosophy, can be complete without careful examination of our best physical theories (this is a much harder task today than it was in the Middle Ages) and that the study of science (and philosophy) itself can be a religious act. I realize, of course, that Maimonides would disapprove, to put it mildly, of my "anthropocentric" conclusions, but this is where my own philosophical quest has led me. But I do hope that this book will also serve to remind religious believers of a Maimonidean truth, the importance of the enterprise of scientific inquiry from a religious point of view.

I hope, finally, that this book can contribute to the dialogue between the sciences and the humanities. I try to show how painstaking attention to what seem to be technical details of mathematical formalism can yield insight into the human mind and its place in nature—a major goal of the humanities. But this insight cannot be obtained, either, without the peculiar skills of philosophical analysis—either mine, or a better one (to paraphrase Newton, if you will forgive the chutzpah).

Of course, from the perspective I adopt here, there is a sense in which physical and mathematical research are the ultimate humanities. I have tried to detail here the extent to which some of the most original scientific ideas in our century were discovered, not by

---

[7] Nothing hangs on this, however—if there is a form of naturalism compatible with anthropocentrism, so much the better. Note, however, that my own anthropocentrism has to do only with the discovery of, not with the content of, present-day theories. Indeed, there may well be a conflict between the two—i.e., scientists use anthropocentric methods, which are doomed to fail, on the very theories they use these methods to discover. I doubt it is profitable to construct a philosophy based both on what scientists say and what they do.

inquiry into nature alone, but also by deploying creatively our own concepts, texts, and formalisms—our "social constructions," to use the current jargon. At the same time, this "research into ourselves" made possible objective achievements—particularly, the discovery of the mathematical structures that govern all aspects of nature. If I am right, then, all participants in the so-called "science wars" saw something valuable, however they misdescribed what they saw.

1

The Semantic Applicability of Mathematics: Frege's Achievements

Many great physicists have expressed amazement that mathematics should be applicable to physics.[1] Eugene Wigner says, "The miracle of the appropriateness of the language of mathematics for the formulation of the laws of physics is a wonderful gift which we neither understand nor deserve."[2] Hertz expressed similar thoughts:

> One cannot escape the feeling that these mathematical formulae have an independent existence and intelligence of their own, that they are wiser than we are, wiser even than their discoverers, that we get more out of them than was originally put into them.[3]

Steven Weinberg:

> It is positively spooky how the physicist finds the mathematician has been there before him or her.[4]

---

[1] I shall mainly speak of applications of mathematics in or to physical science, rather than to the physical world or to the Universe. For I shall try to refrain from taking up a position concerning "scientific realism." Given the variety of positions, however, that answer to the name "realism" (and, for that matter, "antirealism"), I doubt whether I can succeed. At any rate, I do believe that the most interesting questions (if not the answers) concerning mathematical applicability can be stated independently of most issues concerning scientific realism.

[2] Wigner 1967, 237.

[3] Quoted in Dyson 1969, 99. Compare also Feynman: "When you get it right, it is obvious that it is right ... because usually what happens is that more comes out than goes in" (Feynman 1967, 171).

[4] Weinberg 1986.

Richard Feynman:

> I find it quite amazing that it is possible to predict what will happen by mathematics, which is simply following rules which really have nothing to do with the original thing.[5]

Kepler:

> Thus God himself was too kind to remain idle, and began to play the game of signatures, signing his likeness into the world; therefore I chance to think that all nature and the graceful sky are symbolized in the art of geometry.[6]

Finally, Roger Penrose characterizes the applicability of mathematics in physics as a

> profound interplay between the workings of the natural world and the laws of sensitivity of thought—an interplay which, as knowledge and understanding increase, must surely ultimately reveal a yet deeper interdependence of the one upon the other.[7]

These sentiments have been either ignored[8] or dismissed[9] by contemporary philosophers.[10]

It is not that philosophers believe that mathematics is inapplicable, or that there are no philosophical problems associated with mathematical applicability. To the contrary: recent work in the philosophy of mathematics often *cites* the truism that mathematics is applicable—in the sciences, in daily life. For example, an author will

---

[5] Feynman 1967, 171.   [6] Quoted in Dyson 1969, 99.

[7] Penrose 1978, 84.

[8] The second edition of the standard anthology, Benacerraf and Putnam 1984, has not a single article on the applicability of mathematics in the physical sciences. Benacerraf informed me that lack of material was the reason. And though an impressive number of books and articles in the philosophy of mathematics has appeared since 1984, almost none of it deals with our topic. A typical collection of articles, Irvine 1990, contains not one on our topic.

[9] "It is no mystery, therefore, that pure mathematics can so often be applied . . . It is a reasonable hypothesis that pure mathematics in general is so often applicable, because the symbolic structures it studies are all suggested by the natural structures discovered in the flux of things" (Nagel 1979, 194).

[10] Dummett 1991a, as we shall see, is an exception; so is Rescher 1984.

maintain—polemically—that only his or her favorite philosophy can account for the applicability of mathematics.[11]

Again, philosophers and physicists often talk past one another on mathematical applicability. Philosophers concentrate upon the applicability of arithmetic; physicists (or physically-minded mathematicians), upon the "miraculous" appropriateness of matrix algebra[12] or Hilbert spaces[13] for quantum mechanics. The physicists see no difficulty in the applicability of *arithmetic* to the world, and may accuse philosophers who focus upon arithmetic of mathematical ignorance.[14] Philosophers return the compliment.[15] Neither charge is just: philosophers and physicists are speaking of "applications" and "applicability" of mathematics in different ways. There is simply no such thing as "the" problem of mathematical applicability.

Needed, therefore, is a comprehensive philosophical analysis of the application of mathematics, an analysis of:

> What it is to *apply* mathematics;
> What it is for mathematics to be *applicable*;
> What philosophical problems the applicability of mathematics raises;
> What solutions are possible.

I intend this book as a contribution toward that goal, but first, let us look at the work of others.

Speaking for Frege, Michael Dummett,[16] almost alone among contemporary philosophers, analyzes what "applying" mathematics

[11] Such claims are offered by structuralists (e.g., Shapiro 1984); empiricists (Kitcher 1983); and logicists (Frege as interpreted by Dummett 1991a). Wittgenstein castigates philosophers for rendering the application of mathematics, external to mathematics, a mystery: "the application," he says, "must take care of itself" (Wittgenstein 1978, 146). All agree, though, that "mathematics" is "applicable," and that philosophy must come to terms with this.

[12] Wigner 1967.

[13] Cf. Kac and Ulam 1971, 163, quoted in Dummett 1991a, 293.

[14] Cf. Mac Lane 1986, who complains that philosophers have little to say about mathematics beyond that of the third grade.

[15] "It cannot be by a series of miracles that mathematics has such manifold applications; an impression of a miraculous occurrence must betray a misunderstanding of the content of the theory that finds application" (Dummett 1991a, 300; his remarks are directed at, among others, Kac and Ulam).

[16] Dummett 1991a, 256–7.

means. Actually, he analyzes *one* of the many concepts of application in mathematics—that connected with the deductive role of mathematics: In the natural sciences, and in everyday life, mathematical theorems function as premises in deductions, including those which predict observations. To *apply* mathematics (in this deductive sense) is simply to use mathematical premises to effect such deductions.

Now, what philosophical problems arise in applying mathematics this way? Consider the argument:

(1) $7 + 5 = 12$,

(2) There are seven apples on the table.

(3) There are five pears on the table.

(4) No apple is a pear.

(5) Apples and pears are the only fruits on the table.

Hence,

(6) There are exactly twelve fruits on the table.[17]

This argument could predict the result of counting the fruits on the table.[18]

But a *semantical* problem lurks.[19] In the statement '$7 + 5 = 12$' of "pure mathematics," the numeral '7' purports to name a mathematical object, the number 7; but in 'Seven apples were on the table,' the term 'seven' looks like a predicate characterizing the apples. (The latter sentence is what I shall call a "mixed context," because it has both mathematical and nonmathematical vocabulary.) This equivocation destroys the validity of the argument (1)–(6). The philosophical problem is, then, to find an interpretation for all six statements that explains its validity. More generally, the problem is to find a constant interpretation for all contexts—mixed and pure—in which numerical vocabulary appears.

This problem does not stem from a metaphysical "gap" between numbers and fruits, between mathematics and the empirical world. It

---

[17] I have found "fruit" used as a count noun in *Webster's Third International Dictionary*; those unpersuaded will have to substitute "pieces of fruit" for "fruits" here and below.

[18] This statement will be qualified below, but the qualification is not germane here.

[19] It was Carl Posy who defined this problem for me.

arises also when we try to count the roots of an equation by adding the number of real roots to the number of imaginary ones.[20]

Frege addressed this semantical problem, and solved it. His solution, moreover, is independent of his thesis that arithmetic is logic. Numerals are always singular terms—their referents are objects, the numbers.[21] All "mixed" contexts in arithmetic reduce to the form

The number of Fs is m

where "is" means identity. Using the notation of Parsons 1964,[22] we can write:

(7) $NxFx = m.$

Numerical attributions are, in the end, predications, but there are surprises:

- The numerical attribution in (7) is not to a physical object or objects, but to the concept F itself. Thus numerical predication is (at least) second-order predication.[23]

- The numeral 'm' is not the predicate, but only part of it.

To formalize rigorously our deduction about fruits, the following theorem is needed:

(8) $\forall F \forall G (NxFx = m \land NxGx = n \land \neg\exists x(Fx \land Gx)$
        $\rightarrow Nx(Fx \lor Gx) = m + n).$

---

[20] Mill's move of physicalizing arithmetic, therefore, is beside the point, as Frege says (1959, §§ 9 ff.). Frege, of course, argued that Mill's view is wrong even for the applications of arithmetic to observables. See the discussion of Mill below.

[21] One could, naturally, also solve the "semantic" problem of the applicability of mathematics with a theory according to which all numerals are really predicates. See Hellman 1989 for an updated version of this strategy.

[22] Dummett fails to acknowledge such seminal articles as Parsons 1964, Benacerraf 1981, and Boolos 1987—despite the significant overlap between them and Dummett's book. The omission is all the more puzzling in light of Dummett's trenchant criticism (Dummett 1991b, xii) of Americans who ignore outstanding British philosophers like the late Gareth Evans.

[23] Although it is true that, according to Frege, the sentence $NxFx = m$ predicates something of the concept F, this does not mean that it can be written $G(F)$, with F the logical subject of the sentence. (Similarly, the operator d/dx is a higher-order function, but we cannot write it as $D(f)$, with f the "argument" of the derivative operator.) This peculiarity of Frege's semantics will not concern us here, but it is discussed extensively by Dummett 1991a, 87–94.

The theorem demonstrates a connection between addition of natural numbers and disjoint set union.[24] For Frege, (8) is a theorem of pure logic, because the objects m and n are "logical objects"; but for questions of application, we need not decide the status of (8).

The next step is instantiating concepts in (8) for F and G. In our case, we instantiate "apple on the table" and "pear on the table" for 'F' and 'G'. These instantiated concepts, it is important to realize, are not *mathematical* concepts at all; nevertheless, the result is still logically (or mathematically) true.[25] The rest is little more than modus ponens, and the deduction (1)–(6) is formally valid.

The following remarks of Dummett, then, are right on the mark:

> Why does Frege think it necessary, for a mathematical formula to be applied, that it express a thought? Plainly because he takes the application of a mathematical theorem to be an instance of deductive inference. It is possible to make an inference only from a thought (only from a true thought, that is, from a fact, according to Frege): it would be senseless to infer from something that neither was a thought nor expressed one. We do not, of course, call every inference an 'application' of its premises: it is in place to speak of application only when the premises are of much greater generality than the conclusion.
>
> Frege tacitly took the application of a theorem of arithmetic to consist in the instantiation, by specific concepts and relations, of a highly general truth of logic, involving quantification of second or yet higher order: if the specific concepts and relations were mathematical ones, we should have an application within mathematics; if they were empirical ones, we should have an external application. Mathematical theories could not themselves consist solely of logical truths involving only higher-order quantification, since they required reference to mathematical objects . . . When we are concerned with applications, however, the objects of the mathematical theory play a lesser role, or none at all, since we shall now be concerned with the objects of the theory to which the

---

[24] I understand addition in (8) to be defined as the "ancestral" (iteration) of the succession relation for natural numbers; in which case (8) is a highly nontrivial theorem, connecting two different mathematical ideas. If we define addition as simple cardinal addition, then (8) is not quite a definition. Either option was open to Frege in writing the *Grundlagen*, since in his *Begriffschrifft* he had already defined the ancestral.

[25] There is nothing paradoxical in this: the second-order logical theorem $\forall p(p \vee \neg p)$ yields, by instantiating the nonlogical proposition "It is raining," the logical truth "It is raining or it is not raining."

application is being made: application can therefore be regarded as consisting primarily of the instantiation of highly general truths of logic. Evidently, a formalist can allow no place for application so conceived. (Dummett 1991a, 256–7)

* * *

Frege, then, solved the *semantical* problem of the applicability of arithmetic and, at least implicitly, other mathematical theories.

To go by recent philosophical literature, though, the chief problems about mathematical applicability are what we can call the "metaphysical" problems. These problems, we are told, stem from a gap between mathematics and the world, a gap that threatens to make mathematics irrelevant. The reason for my neglect of these "problems" is simple: Frege solved them. However, Frege never emphasized his solution, though it does appear in his *Grundlagen*. Had he done so, the philosophical public would have been spared years of superfluous discussion.

What, exactly, are these "metaphysical" problems?

One is the very existence of mathematical "objects" and mathematical "truths," which some philosophers simply cannot accept. One such theorist is Hartry Field.[26] His view of Frege's project amounts to the following: Frege's—valid—interpretation of arithmetic demands the existence of objects (numbers, sets) that (in Field's view) do not exist. Hence, both the theorems of pure mathematic and the "mixed" propositions of mathematical physics turn out to be *false* statements.

If both pure and "mixed" mathematics were true, there would be no mystery about our reliance on mathematics in making empirical predictions. For what follows from truth by valid logical reasoning is simply true. But if Field is correct, the premises of such derivations are all false—so we need an explanation of how systematically false premises can lead to systematically true conclusions. I refer the reader to Field's writings for enlightenment on this point.

Suppose, however, that we abandon what Dummett (paraphrasing Wittgenstein)[27] calls "the superstitious nominalist horror of

[26] See Field 1980 and Field 1989.
[27] Wittgenstein 1978, 122: "The superstitious dread and veneration by mathematicians in the face of contradiction."

abstract objects in general."[28] That is, suppose we grant the Platonist the existence of mathematical objects. Are there *still* problems—stemming from the *metaphysical* gap between the mathematical and the physical "worlds"—obstructing the use of mathematics in physics? The only one I can envision is this: the "metaphysical gap" blocks any nontrivial relation between mathematical and physical objects, contradicting physics which presupposes such relations.

Surprisingly, Dummett himself endorses this very argument, for he says:

> Some have wished to maintain that [mathematics] is [about] a super-empirical realm of abstract entities, to which we have access by means of an intellectual faculty of intuition analogous to those sensory faculties by means of which we aware of the physical realm. Whereas the empiricist view tied mathematics too closely to certain of its applications, this view, generally labelled "platonist," separates it too widely from them: it leaves it unintelligible how the denizens of this atemporal supra-sensible realm could have any connection with or bearing upon conditions in the temporal, sensible realm that we inhabit.[29]

In Dummett's book on Frege's philosophy of mathematics, the same sentiment occurs:

> Platonism is the doctrine that mathematical theories relate to systems of abstract objects, existing independently of us, and that the statements of those theories are determinately true or false independently of our knowledge. This doctrine . . . raises immediate philosophical problems . . . how can facts about [immaterial objects] have any relevance to the physical universe we inhabit—how, in other words, could a mathematical theory, so understood, be *applied*? (Dummett 1991a, 301)

---

[28] Dummett 1994, 19. On p. 16, Dummett argues that the legitimacy of abstract objects follows from the legitimacy of certain whole sentences which happen to contain terms referring to these objects. (We need not study the direct relationship between the term itself and the object.) The nominalist, then, is to be "pitied" for being in the grip of a "misleading picture."

[29] Dummett 1994, 12, and never explicitly repudiated later on. What is "surprising" and even confusing here is that this argument smacks of just the sort of nominalist "superstitious horror" of abstract objects that Dummett condemns later in the name of Frege.

Underlying these complaints is an argument like this:

(1) On the platonist view, physical laws and theories must express relations between mathematical and nonmathematical objects.
(2) Every relation in physics is a causal (or spatiotemporal) relation.
(3) Mathematical objects do not participate in causal (or spatiotemporal) relations.

Therefore,

(4) On the platonist view, all physical laws and theories are false.

Dummett holds that this, the only argument I can extract from his words, defeats Gödel's (and any other) platonism. He recognizes, of course, that Frege's view is also platonist, but Frege gets off quite lightly:

> [Frege's] combination of logicism with platonism, had it worked, would have afforded so brilliant a solution of the problems of the philosophy of mathematics . . . Frege's idea was that [mathematical] objects should always be defined as extensions of concepts directly related to the application of the mathematical theory concerned: concepts to do with cardinality in the case of the natural numbers . . . In this way, application could be understood as being no more problematic than it would be according to non-platonist logicism: it would not consist in pure instantiation of formulas of higher-order logic, but would involve deductive operations so close to that as to dispel all mystery as to how application was possible. A mathematical theory, on this view, does indeed relate to a system of abstract objects in the sense in which we speak of pure sets . . . they are objects characterized in such a way as to have a direct connection with non-logical concepts related to any one of the particular domains of reality, the physical universe among them. They could not otherwise have the applications they do. (Dummett 1991a, 303)

The truth is, though, that all platonists[30] can benefit from Frege's technical achievement. Frege argued that the laws of arithmetic are

---

[30] Strictly speaking, everyone can benefit from Frege's achievement, but those philosophers who deny that arithmetic statements are true will be faced by additional

second-order laws governing all concepts whatever. Not only did he argue this point, he constructed a deductive system of arithmetic in which this second-order character is evident. In Frege's system, numerals appear in second-order predicates applying to ordinary concepts. In this sense, Frege "predicates" natural numbers of concepts. The concepts themselves may be true of physical objects. In short, mathematical entities relate, not directly to the physical world, but to concepts; and (some) concepts, obviously, apply to physical objects. The mystery thus vanishes without a trace. As Frege put it himself in the *Grundlagen*, "The laws of number, therefore, are not really applicable to external things; they are not laws of nature. They are, however, applicable to judgments holding good of things in the external world: they are laws of the laws of nature" (Frege 1959, § 87).

This disposal of the "metaphysical" problem of the applicability of arithmetic to the physical world depends not at all upon Frege's logicism. For example, suppose we regard set theory, rather than second-order logic, as the foundation of all mathematics, because all classical mathematics can be modeled in it. Frege's insight adapts readily to this new context: numbers characterize sets, not physical objects; while sets can contain, of course, physical bodies. Set theory is applicable, in the present sense (one of many senses, I remind you), simply because physical objects can be members of sets. This is a thoroughly nonmystical idea, always supposing we accept the existence of sets in the first place.

Even the inconsistency of Frege's logical system (the one of the *Grundgesetze*) does not mar Frege's solution of the metaphysical problem of applicability. As George Boolos has shown, the program of Frege's *Grundlagen*, including all theorems there sketched, goes through in a consistent second-order theory,[31] which he calls "FA" (Frege Arithmetic), in which the only "nonlogical" axiom is

---

problems concerning mathematical applicability. Field is one such philosopher, since he takes arithmetic propositions literally as implying the existence of numbers, yet he denies that numbers exist. Therefore, he benefits from Frege's program in that Frege shows how to incorporate pure and mixed mathematical statements in a single deduction (something that he would have had to do), but he remains with a problem not facing Frege: how systematically false statements allow systematically true empirical predictions.

[31] Other philosophers have asserted this point, but Boolos not only constructed FA, but proved its consistency. Hence the attribution to Boolos, of whose work Dummett seems unaware.

$$\forall F \exists! x \forall G (G \eta x \leftrightarrow F \, eq \, G).$$

(The eta sign replaces the usual epsilon set membership, and expresses the relation that holds between a concept G and the extension of a higher-level concept under which G falls. The sign "eq" expresses the equinumerosity of the concepts.) That is to say, the only mathematical objects Frege needs for arithmetic are classes of equinumerous concepts. Whether we call FA "logic" or not is here irrelevant: FA captures the benefits of Frege's approach to arithmetic, logic or no.

Nor is this insight limited to arithmetic, since mathematicians have modeled all classical mathematics in set theory, ZFC. To "apply" set theory to physics, one need only add special functions from physical to mathematical objects (such as the real numbers). Functions themselves can be sets (ordered pairs, in fact). As a result, modern—Fregean—logic shows that the only relation between a physical and a mathematical object we need recognize is that of set membership. And I take it that this relation poses no problems—over and beyond any problems connected with the actual existence of sets themselves.

We can now conclude that Frege completely solved the semantic and the metaphysical problems of applicability.

In the *Grundlagen*, Frege showed how to interpret both pure and mixed arithmetic statements so that we can use pure mathematics to deduce "applied" conclusions. This solves the semantic problem. He did not specify the underlying logic, but all of his proofs can be codified in Boolos's FA word for word. (That FA is not "logic" is irrelevant to the semantic problem of applicability.)

And, in solving the semantic problem, Frege did not need to postulate any metaphysically suspect relations (such as causal relations) between mathematical and nonmathematical objects. Mathematical objects are related only to other mathematical objects and to concepts. That physical objects may fall under concepts and be members of sets is a problem only for those who do not believe in the existence of concepts or sets. Perhaps without even intending to, Frege disposed of the metaphysical "problem" of applicability, and rendered superfluous most recent discussions of "the" problem of applicability.

2

The Descriptive Applicability of Mathematics

Can we conclude that Frege solved *every* problem concerning the applicability of arithmetic? No. Frege left other problems untouched, and these will be my problems.

It is crucial to distinguish problems concerning specific mathematical concepts from those having to do with mathematical concepts in general. To postpone the chore of actually characterizing the mathematical concepts,[1] I begin with the former problems. And the specific concepts I begin with are, again, the arithmetical ones: addition and multiplication.

For example: what makes arithmetic so *useful* in daily life? Why can we use it to predict whether I will have carfare after I buy the newspaper? Can we say—in *non*mathematical terms[2]—what the *world*[3] must be like in order that valid arithmetic deductions should be effective in predicting observations?[4]

[1] See Chapter 3.

[2] This requirement is not inspired by the project of nominalizing physical theory of Field 1980. I am not interested in translating any physical theory into a nominalistic language, but explaining, in nominalistic language, the conditions under which a mathematical concept will be applicable in description.

[3] I realize this five-letter word offends some philosophers; they can paraphrase it out of the next few paragraphs. Most of the problems concerning the applicability of mathematics in natural science and in daily life cut across the realist/antirealist divide, I would like to believe. On the other hand, the *solutions* to the problems may well be sensitive to the realist/antirealist controversy.

[4] To put the matter in Kant's language (with thanks to Carl Posy): what is the "objective validity" of such logical or mathematical concepts as disjoint union, cardinal number, etc.? Cf. also the introduction to Detlefsen 1992, where the editor explicitly draws this parallel.

These were not Frege's questions, and could not have been: he attended to the applicability of mathematics in general, not to nature specifically. His concern was not with the usefulness of mathematical reasoning, but its validity—to which the state of the world is immaterial.

Frege treats the *semantical* applicability of mathematical *theorems*; I will attend to the *descriptive* applicability—the appropriateness of (specific) mathematical *concepts* in describing *and lawfully predicting* physical phenomena.[5] Whereas, for Frege, applying meant "deducing by means of," for me it will be "describing by means of."

Despite this, Frege continues to guide us. Numbers, according to his theory, take the measure of concepts; concepts qualify objects. Concepts here are the *physical* concepts, those applying to physical objects (below: bodies).

Consider now a predicate P, for example "coin in my pocket." Whether an object is a P—indeed, whether it exists—can change over time. Speaking with the vulgar, we say that the extension of predicates can also change over time, i.e., that sets can change, a confused if intuitive way of talking. Thinking with the learned, we follow Quine: physical predicates apply, timelessly, to "time slices" of bodies.[6] It is the wax-at-t that is soft, not the wax. When we speak of the number of objects of type P at a time, we mean the number of object-slices-at-t of type P. To say of a predicate that its extension changes over time means that the predicate is true of t-slices of different objects for different times t. If the extension of a predicate changes too rapidly at t (speaking again with the vulgar), then humans cannot discern the number of Ps at t. Arithmetic—a technique for inferring the number of objects in one set from the number in others—will then be useless, though "true." Arithmetic is useful because bodies belong to reasonably stable families,[7] such as are important in science

---

[5] The descriptions of which I speak are thus *lawlike* or *projectible* descriptions in the sense of Goodman 1983: descriptions which could appear in natural laws and thus be used in predicting events. Only these descriptions are my concern.

[6] Quine 1960, § 36.

[7] I readily grant that if this condition were not met, human experience would be impossible, not only arithmetic. But this would at most show that the condition is an *a priori* one. I agree with Kripke that Kant erred in thinking that every *a priori* truth is a necessary truth.

and everyday life.[8] The number of coins in my pocket; the number of fruits on the table; the number of political parties even in Israel; all stay constant long enough for humans to count them. The invariance is maintained, not only through time, but under translations and other common maneuvers. The coins in my pocket are usually the same whether or not I walk around the house, put candies in my pocket, too, and so forth.

Another factor curbing our ability to count a set of objects, aside from the above-mentioned volatility of its members, is their dispersion. Define the dispersion of a set of bodies as the average distance between any pair of its elements. Generally, a physical predicate is more significant to us arithmetically if its extension has low dispersion. (By *abus de langage*, the dispersion of a predicate will mean the dispersion of its extension.)

Now in many cases of interest, as a result of performing physical operations (pushing, pulling) at time t on the set of Ps (for example "coins belonging to me at time t"), it becomes possible to define a predicate Q ("coins in the basket at t'") which has the same extension at t' that P had at t, but is much more concentrated (has much smaller dispersion). I will call predicate Q a *time-aggregation* of P, relative to times t and t'. It is a contingent result of human intervention that time-aggregations of predicates often exist.

It is contingent,[9] too, that when the elements of a set are pushed and pulled around, they often retain some of the properties which interest us. Coins remain coins, scattered or clustered.

Consider now how addition is applied, descriptively, to events. John Stuart Mill explained addition by what he thought of as its paradigm application: Suppose I throw five pennies into an empty hat and then four more. Most likely, the number of pennies in the hat is now nine. The identity $5 + 4 = 9$ elevates this empirical prediction to a general law. Frege retorted that Mill confuses the meaning of addition with its application.[10] In other words, we apply addition to

---

[8] True, families lacking stability of this kind could not play a role in daily life in the first place. But it is contingent that there are any such families.

[9] I say contingent, not empirical, because it seems reasonable that those properties of interest to human beings would be those invariant under these kinds of transformations. Even if *a priori*, though, the fact is contingent—again, just as Kripke pointed out.

[10] Frege 1959, § 9.

assembling, gathering—without interpreting '+' *as* an idealized assembling. The sign '+' means the same in every context.

Frege's criticism is too tame: Mill's example does not even *apply* addition to "gathering." Since the idea that "gathering" is an application of addition seems widespread, it is worth refuting.

Given A and B, respectively the disjoint extensions of concepts F and G at a time, if there are m objects in A and n in B, then there are $m + n$ objects in $A \cup B$, i.e., $\{x: Fx \lor Gx\}$, as a matter of logic, or set theory, certainly not physics.

Suppose that F and G have low dispersion, as arithmetically useful predicates tend to do. The disjunction of F and G will have greater dispersion than either F or G alone if the Fs and the Gs are in different places. In no case will the dispersion go down. Using jargon, *dispersion increases monotonically under set union* (predicate disjunction). Consequently, disjunctive predicates are often less arithmetically useful than either disjunct; why is addition useful where it enumerates an uninteresting set?

That we can neutralize dispersion makes addition useful: a time aggregation of the "uninteresting" $F \lor G$ then exists. The same bodies, dispersed now, are concentrated later—by human effort. What links addition and gathering, then, is this: gathering neutralizes the dispersion that set union so often entails.

To counter dispersion by force is to leave the compass of logic. Addition is useful because of a *physical* regularity: gathering preserves the existence, the identity, and (what we call) the major properties, of assembled bodies. But then *gathering* is not an application of addition at all.[11]

---

[11] What, then, is being applied in the "caveman" arithmetic of sticks and stones? Consider a number of stones. We can distinguish between (a) the (scattered) physical object X of which the stones are parts; and (b) the set Y of the stones, i.e., the abstract collection of which the stones are *members*. The physical object (a) is called the "mereological sum" of the stones. Now, it is a physical fact that the scattered physical object X retains the same stones as its parts under a wide range of physical changes, such as gathering the stones together. The parts of X which are stones, therefore, are members of the same set Y over time (thinking with the vulgar). Thus the invariant structure of the object X is the set Y. The conclusion is that in the arithmetic of sticks and stones, the concept of a mereological sum is an application of the concept of a set. This is not too surprising, since the concept of a set is an abstraction from that of a mereological sum. But since most people find it difficult to grasp clearly the concepts of set and of sum, the applications here are not usually made consciously. That is, it is possible to

Contrast *weighing*, a true application of addition. If one body balances 5 unit weights, and another balances 4, then both together will usually balance $5 + 4 = 9$ unit weights. The natural numbers indirectly describe, by laws of nature, not only the sets of unit weights placed on the scale, but the objects they balance. Addition of numbers becomes a metaphor for "adding" another object to the scale. Arithmetic is not empirical, but it predicts experience indirectly by the law: if m and n are the numbers of unit weights that balance two bodies separately, then $m + n$ units balance both.[12] Equivalently: if one object weighs m units, and another weighs n units, then the (mereological)[13] sum of both "weighs $m + n$ units." This more usual expression looks like a tautology, but is as empirical as the former:[14] the expression '$m + n$' is embedded in a nomological description of a phenomenon (weight). This description induces an isomorphism between the additive structure of the natural numbers and that of the magnitude, weight.

In referring to an "additive structure," I do *not* mean a system of bodies. There are too few bodies for them to correlate with all the natural numbers. The isomorphism is between the natural numbers and a *magnitude*: infinitely many physical properties parametrized by those numbers. As with every magnitude, not all of those properties need actually materialize.

Now let's examine multiplication. The paradigmatic operation here is, ostensibly, *arranging*: in equal rows, equal groups, etc. This is how we teach multiplication to children. But here, too, arranging is not an application of multiplication; rather, arranging makes multiplication valuable. As concentrating a *dispersed* set makes it useful, so does arranging a *large* set make it "surveyable."[15] Multiplication enumerates the union of (pairwise disjoint) similar sets, but arranging the elements of the sets in rows allows us to grasp that number.

---

apply a mathematical concept without being aware one is doing so, just as it is possible to speak prose one's whole life without realizing it.

[12] Actually, according to Einstein's theory of general relativity, weight is not additive, as I shall discuss below. On the human scale, the deviation is experimentally undetectable.

[13] The mereological sum of a number of bodies is the one scattered body with the same molecules as the bodies.

[14] A fascinating study of how empirical laws come to pose as tautologies is Levy-Leblonde 1979.   [15] This term is from Wittgenstein 1978.

That we can arrange a set without losing members is an empirical precondition of the effectiveness of multiplication, rather than one of its applications.

A familiar and genuine application of multiplication is tiling with unit squares. Suppose we have a rectangular floor and we inquire how many tiles cover it. The elementary answer is that if the floor length is m units and the width is n units, the number needed is usually $m \cdot n$. As in weighing, we have an isomorphism. The numbers m and n come to measure, not just the size of a set (of units), but the length of lines. Multiplication comes to portray decomposing the rectangle into squares by parallel lines; conversely, moving from one-dimensional to two-dimensional Euclidean "intervals." (Historically, of course, it is the other way around—multiplication was seen as this very geometric operation; and thus the product of the two numbers was seen as a different sort of number from the multipliers. We cannot speak of an "application" of multiplication to geometry until the modern period.) That $m \cdot n$ counts those squares is an empirically[16] based isomorphism of the multiplicative structure of the natural numbers with the two-dimensional geometrical structure of the plane.

The use of addition in weighing, of multiplication in tiling, involve paradigmatic, indeed prehistoric, activities: gathering, arranging. But these activities are not ends: in gathering weights into the scale pan, we balance another object; in arranging the tiles, we cover a floor. Gathering and arranging are so linked with weighing and tiling that the additive or multiplicative structures of the natural numbers characterize nonarithmetic structures, too.

Is there anything unexpected about the descriptive usefulness of addition and multiplication? No; it is not hard to set down conditions, in nonmathematical language, for a magnitude to have an additive structure. Indeed, the theory of measurement sets forth the conditions under which a magnitude has the additive structure of the reals. It is clear that we need not adopt Mill's "empiricist" position on arithmetic[17] to explain the descriptive applicability of the arithmetic operations.

---

[16] I am assuming here that geometry is empirical because I think it is. If you think it isn't, please ignore this example.

[17] An empiricist need not, and most do not, hold that *mathematics* is empirical. Mill is the exception.

* * *

Consider now *linearity*: why does it pervade physical laws? Because the sum of two solutions of a (homogeneous) linear equation is again a solution. This property corresponds to the Principle of Superposition, exploited by Galileo: joint causes operate each as though the others were not present. If we shoot a cannonball directly up, its motion is the sum of a constant (inertial) rising produced by the cannon, and an accelerated falling caused by gravity. The position of the cannonball is thus given by an algebraic sum (in general, the vector sum) of the two displacements.

Or consider the classical or quantum electromagnetic field, meaning the electromagnetic field without "sources," i.e., charged particles. This theory is linear, which means that the various electromagnetic waves which compose the field do not interact with one another, in accordance with the Principle of Superposition. (If there is even one charged particle, it interacts with the electromagnetic field as it moves around, and the equation is no longer linear.) In quantum mechanics, of course, the electromagnetic waves are identified with particles known as photons. So our conclusion is that photons do not interact with one another to change the field, which is to say that the photons themselves carry no electric charge.

By contrast, Einstein's theory of gravity has a small nonlinear effect, and the physical meaning of this is the reverse: the hypothesized particles of the gravitational field, the "gravitons," do interact with one another, or carry "gravitational charge," the same as mass. Similarly for the nonlinear theory of the nuclear force field. The hypothesized particles composing this field are called "gluons" and they, too, interact, because they carry the same kind of "nuclear charge" as do the protons and neutrons themselves.

Both linearity and nonlinearity, then, have a clear physical correlate, based on superposition; all is explained—so far. But we must investigate another role for linearity in science.

Obviously, not every equation in science is linear. The planets travel in ellipses, for example, not straight lines, even though the fundamental equation $F = ma$ is linear. Even so, linearity retains considerable importance, since the nonlinear may often be approximated by the linear. For example, we approximate a curve, over short distances, by its tangent, an idea which finds full flower in the famous

Taylor series expansion for functions. Approximations like this are valid if the curve is smooth, or at least has smooth pieces, certainly a physical property. Hence we have an explanation for the second role of linearity in science: the smoothness of many natural processes. (Even in quantum mechanics, the wave function evolves smoothly through time.)

From this, it follows that where nature does not operate smoothly, linearity loses application. And, indeed, exponents of "fractal geometry" like Benoit Mandelbrot argue that nature is best described by infinitely rough, not (piecewise) smooth, curves. An example is Figure 1: a picture of a "fern," generated by a fractal-producing computer program.[18] Even smooth configurations, they say, can evolve

FIGURE 1

toward roughness. An instance is soot particles in a colloid that stick one to another and grow into a fractal pattern (Mandelbrot and Evertsz 1990).

[18] See, for example, Mandelbrot 1990 and the other essays in Fleischmann et al. 1990.

If fractal, rather than smooth, geometry is what describes nature, the physical implications are immediate: gone are the prospects for understanding phenomena by breaking them up into their component parts. The mathematical device for doing this is the Taylor series, and the method works only if the functions approximated are smooth. Mandelbrot therefore advocates a complete revision in scientific thought, away from understanding the whole in terms of its parts.

Mandelbrot may well *be* wrong,[19] but he cannot be *proved* wrong *a priori*. Linearity is applicable to the extent, and only to the extent, that the Principle of Superposition holds, and to the extent that nature operates in a smooth, or at least piecewise smooth, manner. Whatever we are to say about this question, we can at least conclude this: there is no mystery concerning the applicability of linearity; the mathematical property of linearity can be reduced to physical properties which nature may either exhibit or not exhibit.

* * *

Scientists have succeeded in explaining the applicability of far more arcane mathematical concepts than linearity, by matching mathematical to physical concepts. One of the most abstract mathematical concepts is known as a fiber bundle, so difficult that one of the greatest physicists of all time, Yang, confessed that he could not readily master the field (I hope, therefore, I will be forgiven if I do not explain it here). Ironically, the remarkable applicability of fiber bundle theory to physics rests on the translatability of the concepts of fiber bundle theory into the concepts of gauge field theory (cf. Chapter 6)—a theory that Yang himself did more than anybody else to discover.

Consider Table I, drawn up by Yang himself (Zhang 1993, 17), showing that (almost) every basic concept from bundle theory has an exact translation into the gauge field terminology. Even the reader who understands not a single word of either terminology (most of my readers, I hope) should be impressed by the detailed correspondence of two independently conceived theories, one physical, the other mathematical. (Of course, only the experts will be able to verify the

[19] Cf. Shenker 1994.

TABLE I. TRANSLATION OF TERMINOLOGIES

Gauge field terminology	Bundle terminology
gauge (or global gauge)	principal coordinate bundle
gauge type	principal fiber bundle
gauge potential b_μ^k	connection on a principal fiber bundle
S_{ba}	transition function
phase factor Φ_{QP}	parallel displacement
field strength $f_{\mu\nu}^k$	curvature
source J_μ^K	?
electromagnetism	connection on a $U_1(1)$ bundle
isotopic spin gauge field	connection on a SU_2 bundle
Dirac's monopole quantization	classification of $U_1(1)$ bundle according to first Chern class
electromagnetism without monopole	connection on a trivial $U_1(1)$ bundle
electromagnetism with monopole	connection on a nontrivial $U_1(1)$ bundle

accuracy of the table.) The philosophical point can be grasped easily; the abstract nature of a mathematical concept is no bar to its reduction to the physical. In short, the applicability of the fiber bundle concept is based on the existence of gauge fields.[20]

Fiber bundles, however, seem to have played no role in the discovery of the gauge field concept. Had they played such a role, the Yang–Mills discovery would still have been mysterious—it would then have been a thoroughly "Pythagorean" discovery, because the "geometry" in question is not spacetime geometry at all. As Yang put it much later:

That non-Abelian gauge fields are conceptually identical to ideas in the beautiful theory of fiber bundles, developed by mathematicians *without reference to the physical world*, was a great marvel to me. In 1975, I discussed my feelings with Chern [the famous mathematician], and said "This is both thrilling and puzzling, since you mathematicians dreamed up these concepts out of nowhere." He immediately protested, "No, no,

---

[20] See Chapter 6 for a detailed discussion of the discovery and Yang and Mills.

these concepts were not dreamed up. They were natural and real."(Yang 1977, reprinted in Yang 1983, 530)

By "real," Chern did not mean "physically real," of course, but "mathematically real," a concept discussed in Descartes' Meditation IV under the name "true and immutable."

Historically, however, geometry played no role—but rather "physics." And I argue in the text that the role of "quantization" in the discovery turns the physics into formal manipulation. Thus, even for Pythagoreans, the success of Yang–Mills is a cause for wonder.

Yang explicitly denies that geometry had anything to do with his and Mills' discovery: "What Mills and I were doing in 1954 was generalizing Maxwell's theory. We knew of no geometrical meaning of Maxwell's theory, and we were not looking in that direction" (Yang 1983, 74). Mathematicians have a hard time taking this denial at face value, especially when they note that the theory of "fiber bundles" was in the air during the fifties, that Yang's father was a mathematician, and, particularly, that the equations of Yang and Mills are truly identical to those of fiber bundle theory. (Cf. Drechsler and Mayer 1977, 2: "A reading of the Yang–Mills paper shows that the geometric meaning of the gauge potentials *must have been clear to the authors*" (italics mine)—quoted by Yang himself, 1983, 74, who says they are simply mistaken.)

But aside from the explicit denials of Yang, we have his candid statement (Yang 1983, 73):

> With an appreciation of the geometrical meaning of gauge fields, I consulted Jim Simons, a distinguished differential geometer, who was then the Chairman of the Mathematics Department at Stony Brook. He said gauge fields must be related to connections on fibre bundles. I then tried to understand fibre-bundle theory from such books as Steenrod's *The Topology of Fibre Bundles*, but learned nothing. The language of modern mathematics is too cold and abstract for a physicist. *All of this happened from 1967 to 1969.* (Yang 1983, 73, italics mine)

Thus, fifteen years after the appearance of the Yang–Mills paper, its principal author, even after becoming convinced of the appropriateness of the geometrical description of the gauge field in terms of fiber bundles, was still "unable" to grasp the theory.

* * *

So far we have seen examples in which the descriptive applicability of a mathematical concept is reasonable and no mystery, in terms of general physical properties of nature. Wherever the property holds, the mathematical concept is there applicable, and vice versa. A concept whose descriptive applicability, though unquestioned, requires elaboration, is that of the inverse square. There are inverse square laws ruling gravity (Newton's law), electrostatics (Coulomb's law), and optics (the intensity of a spherical light wave). The experimental accuracy of these laws—for gravity, more than one in ten thousand as of 1910—particularly impresses Eugene Wigner (Wigner 1967). The explanatory challenge, then, is to explain, not the law of gravity by itself, but the prevalence of the inverse square.

This question much exercised the great American philosopher-scientist, Charles Peirce:

> 7.509. But meantime our scientific curiosity is stimulated to the highest degree by the very remarkable relations which we discover between the different laws of nature,—relations which cry out for rational explanation. That the intensity of light should vary inversely as the square of the distance, is easily understood, although not in that superficial way in which the elementary books explain it, as if it were a mere question of the same thing being spread over a larger and larger surface....But...what an extraordinary fact it is that the force of gravitation should vary according to the same law! When both have a law which appeals to our reasons as so extraordinarily simple, it would seem that there must be some reason for it. Gravitation is certainly not spread out on thinner and thinner surfaces. If anything is so spread it is the potential energy of gravitation. Now that varies not as the inverse square but simply [as] the distance. Then electricity repels itself according to the very same formula....Here is a fluid repelling itself but not at all as a gas seems to repel itself, but following that same law of the inverse square.

> 7.510. According to the strictest principles of logic these relations call for explanation . . . you must explain these laws altogether.[21]

What Peirce is looking for is some general physical property which lies behind the inverse square law, just as the principle of superposi-

[21] All quotes are from Peirce 1958.

tion and the principle of smoothness lie behind linearity. Peirce rejects as the explanation the obvious property of Euclidean space that the surface area of a sphere is proportional to the square of the radius (for the sake of argument, regarding geometry as though it were physics).[22] He is therefore left without any explanation of the applicability of the inverse square law, though he does feel that there ought to be one.

* * *

I shall now cite a number of examples of mathematical concepts whose descriptive applicability *now* seems mysterious.

Consider, first, the applicability of complex analysis in physics; in particular, that of an analytic function. A function of a complex variable is said to be *analytic* in a region of the complex plane if it is differentiable everywhere in the region. Now, the differentiability of a function of a complex variable is a much stronger condition than the differentiability of a real variable. In fact, for a function of a complex variable to be differentiable, it must satisfy special equations, the Cauchy–Riemann equations, which functions of a real variable do not have to satisfy.[23] Furthermore, it is a theorem that if a function of a complex variable is differentiable once, it is differentiable infinitely many times—again, this is not true of functions of a real variable. From the latter theorem we derive that a function of a complex variable is analytic if, and only if, it can be represented by an infinite power series (Taylor series), since the conditions for such representation by a power series are precisely that all the derivatives of the func-

---

[22] This explanation, by the way, was published earlier by Immanuel Kant in his *Prolegomena to Any Future Metaphysics* (Kant 1950), sec. 38.

[23] Differentiability for a complex function is defined, verbally, the same way as for a real function. However, since we are dealing with the complex *plane*, this limit must stay the same no matter by which of the infinitely many paths h goes to zero. Thus, differentiability for a complex function is a much stronger condition than for real functions, and this condition is expressed by the Cauchy–Riemann equations: writing $z = x + yi$ and $f(z) = u(x,y) + iv(x,y)$, where u and v are real-valued functions of x and y, then a necessary condtion for $f(z)$ to be differentiable in a domain D is that the Cauchy–Riemann equations

$$\frac{\partial u}{\partial x} = \frac{\partial v}{\partial y}, \; \frac{\partial u}{\partial y} = -\frac{\partial v}{\partial x}$$

hold at each point of D; where, of course, all four partial derivatives are assumed to exist everywhere in D.

tion exist. (For a function of a real variable to be representable by a power series, more than infinite differentiability is needed.)

The concept of analyticity turns out to be astonishingly applicable. Let's look at three examples: to fluid dynamics, to relativistic field theory, and to thermodynamics.

The applicability of analyticity to hydrodynamics follows from the theorem that a two-dimensional ideal fluid (i.e., one in which the third dimension plays no role in the problem) obeys the Cauchy–Riemann equation. This makes the theorems of analytic function theory immediately applicable to ideal fluids.

The applicability of analyticity to relativistic field theory follows from theorems that link functions defined on a light cone to analyticity.[24]

Finally, an application of analyticity to thermodynamics is the assumption that one can treat the critical temperature of a ferromagnet (the temperature at which it loses its magnetism) as an analytic function of the number of dimensions of the magnet.

Now, even if we were to regard the first two applications (hydrodynamics and relativistic field theory) as sufficiently explained by the theorems quoted, note that there is no one physical property which explains all three applications, or types of application. So the situation does not resemble the case of additivity, where one property explains just about every application of "+" in physics.

In addition, the first application, to hydrodynamics, is true only for two-dimensional ideal fluids, in keeping with the essential two-dimensionality of the theory of functions of a complex variable. Thus, there is something "accidental" about the applicability.[25]

[24] Consider a "light cone" in spacetime. Outside the light cone, we can say that a particle has zero probability to be found; in mathematical terminology, the position function has "support" in the cone. Now there are a number of theorems relating (certain) "supported functions" to the analyticity of their Fourier transforms; for details, see Reed and Simon 1975, ix.3, especially Theorem IX.16. (The functions in question decrease suitably rapidly for large arguments, and have nice smoothness properties.)

But, in quantum mechanics, the momentum function is the Fourier transform of the position of function; and the notion of a light cone is characteristic of relativity theory. Thus, in the context of relativistic quantum (field) theory, the concept of analyticity may have a physical interpretation. (See Wightman 1969 for a popular discussion of this point.)

I am grateful to Barry Simon and Larry Zalcman for help here.

[25] I am grateful to Joel Gersten for pointing this out to me.

The third application, though, is totally mysterious, from the point of view of explaining the descriptive applicability of analyticity. The assumption that the critical temperature of a magnet is an analytic function of its dimension is, in fact, physically meaningless. Not only will we have to condone, in physics, magnets of dimension 3.5 (we have gotten used to such things by reading about fractals), but we will have to swallow magnets of dimension $2 + 3i$! Here the analytic function is used as a calculational tool or formal trick: we cannot calculate the problem for three dimensions, so we calculate it for a four-dimensional magnet, then expand the function as a power series in the complex plane around the number 4, and plug in the value 3. Nobody knows (today) why this works.

Regarding analyticity and its descriptive application, we can sum up the situation as follows. We understand some individual applications, some much better than others; we understand some applications not at all; and in any case, we have no one property that corresponds to analyticity in all applications.

* * *

We return now to where we began—the mathematical foundations of quantum mechanics. To see why Kac and Ulam regard the appropriateness of the Hilbert space concept in quantum mechanics as a miracle, we must go deeper than Dummett has done. I will give an overview of the puzzling role of the Hilbert space concept in quantum mechanics in this chapter; for a more detailed treatment of the same material, see Appendix A (there, I provide a "formal" derivation of quantum mechanics, including the concept of "spin").

The descriptive applicability of Hilbert space to quantum mechanics follows from what I shall call the "maximality principle." About this, a few words.

A Hilbert space is a kind of vector space, and it is the vector space concept which is the heart of the mathematical formalism of quantum mechanics. A central concept is that of a basis for a vector space, which we can think of intuitively as a set of "axes" as in Cartesian geometry. The number of axes is the dimension of the space. A vector is represented by a point in the space. Each vector in an n-dimensional space, relative to a basis, has n "coordinates." (In a Hilbert space, the number of coordinates can be infinite.) A crucial point is that we are free to choose any basis, any set of axes, we want, so that the coordinates of a vector will always be relative to the chosen basis.

In quantum mechanics, a physical system is always described by a vector space, and the state of the system, by a *unit vector* in that space. The evolution of the system through time corresponds to the "rotation"[26] of the vector in different "directions" under the influence of the various forces of nature.

In order for the vector space to make numerical predictions, one must be able to say, at any moment, what the coordinates of the unit vector are. In other words, one must choose a set of basis vectors, or "axes," onto which we can project the unit vector to get numbers. The Maximality Principle has to do with the choice of basis, and goes as follows, in a deliberately unrigorous treatment.

If a Hilbert space H represents a quantum system Q, then each basis, or set of "axes," of H corresponds to a physical property of Q; and each physical property of Q corresponds to a basis, or at least a subset of a basis, of H (Maximality Principle).[27]

In particular, the magnitude of *position* corresponds to a complete basis. Thus, position information about a system *at a given time* determines information concerning every other magnitude of the system *at that time*. This information is obtained simply by changing the basis of the Hilbert space, and recalculating the coordinates of the unit vector relative to the new basis.

This principle has no correlate in classical mechanics, and is, at least today, physically unintelligible (meaning, as usual, that there is no other language to express the principle). It is not just that position determines the other properties—but the way that this determination takes place.

Let us see how it works. Consider a single particle, represented by a unit vector. Choose the "position basis." Then we have infinitely many coordinates for the vector, because there are infinitely many places the particle could be. When we square these coordinates, we get the probabilities that the particle will be found in each of these

---

[26] Technically, of course, I am referring to a unitary transformation of the vector.

[27] Some physicists would accept only the latter clause, because they claim that there are some bases that correspond to "unphysical" properties, needing to be weeded out by "superselection rules."

places—a probability distribution.[28] This is as much information about the position of the particle as quantum mechanics allows. If we now desire information about some other property of the particle, such as its momentum, we simply change the basis to get a new set of coordinates. That is, we consider the same unit vector relative to a new set of "axes." The same for angular momentum and the like. The only thing that remains, then, is to "locate" these different bases— those of linear momentum and angular momentum—and it turns out to be a rather simple exercise to do this (cf. Appendix A). That is, *given* the strange idea that each property of a system determines every other one, and this by a change of basis—which we can think of also as a "rotation" of the axes to a new position—simple arguments show that, for example, the "momentum axes" are inalterably fixed relative to the "position axes." And so for the other properties with their "axes."

The Maximality Principle, though certainly not physically intelligible, is a contingent, falsifiable proposition. In fact, it is such a strong constraint on nature that it barely escapes inconsistency.

For example, moving our measuring device 100 meters away will change every position measurement. So the unit vector will have to "move" to a different place, relative to the position "axes." But such a translation will not change any momentum measurements. How can a unit vector move so that the position information it imparts (concerning particles) changes, yet the momentum information (concerning those very particles) stays the same? How could some, but not all, of the changes in a unit vector be physically meaningless?

The puzzling answer: if the coordinates of the unit vector are *complex numbers*—which have not only magnitude, but also *phase* ("direction")—the unit vector can accomplish its "impossible" mission. The key idea is: the magnitudes of the coordinates are physically meaningful, not their phase.

Strangely, the addition of these physically meaningless quantities to physics allows just the extra degree of freedom the unit vector needs. For each physical property, we associate a different set of coordinates on the *same* space to the *same* unit vector. Thus we have

---

[28] Squaring the coordinates ensures that the probabilities will be positive numbers, as they must be.

momentum coordinates, position coordinates, etc. These coordinates give information concerning momentum, position, etc. The coordinates are complex numbers. To change the phases of some or all the momentum coordinates, leaving the magnitudes alone, is to convey the same momentum information. Yet this very change of phases will almost always cause the *magnitudes* of *other* sets of coordinates to change, for example position coordinates. In a more familiar vocabulary, the Maximality Principle leads to the wave-particle duality, to interference phenomena, etc.

But these phenomena arise only when we make a *nonuniform* change of phases. Consider one particle with a fixed momentum. One momentum coordinate will be enough to determine its state. Only one momentum coordinate, that is, will be nonzero. Suppose we move our measuring device as before. Only the phase of this single momentum coordinate can change, which amounts to a *uniform* change of phase[29] that cannot lead to interference effects. None of the magnitudes of the position coordinates can change. This means that moving the measuring device effects no change in the position information that the device gives us. This is absurd, unless the device gave us zero information in the first place. We thus arrive at the sensational result that exact information concerning momentum wipes out all information concerning position—a special case of the Heisenberg Uncertainty Principle! All this follows from a formal premise, the Maximality Principle, which does not correspond to any physical idea.

Momentum is not the end of the matter. There are other observables in physics—angular momentum, for example. The Maximality Principle demands that the state vector give us information about this too. The consequences are startling: in order for angular momentum to be in the same Hilbert space as the other quantities, it must[30] be quantized in integral multiples of a minimum![31] We arrive at the "quantum" principle, which physics contended with for a quarter of a century, free of charge.

But not only is angular momentum quantized, its *direction* is quantized. This means that, relative to some axis—say the z-axis—

[29] A zero coordinate remains zero under any change of phase.
[30] Given reasonable mathematical conditions on the state vector, outlined below.
[31] For details, consult Appendix A below.

there are only a finite number of directions for the "orbit" of a particle. If the angular momentum of an orbiting particle is fixed at j units, and the angular momentum is conserved, we can think of it as a unit vector inhabiting, not a whole Hilbert space, but a finite-dimensional vector space,[32] one dimension for each direction of the orbit. (Of course, we are ignoring, for the purpose of the example, every other property of the particle.) As we rotate the laboratory in different directions in space, the unit vector moves around in the finite-dimensional space, changing the probabilities that we shall find the orbit of the particle in one of the allowed spatial directions. What I have described in lay terminology is what the mathematicians call an *irreducible representation of the group SO(3) of spatial rotations*.

That the various groups of transformations such as rotations act not on Euclidean space, but on a linear (vector) space, puts enormous constraints on quantum mechanics, and therefore on nature. Here is an example: it is a theorem that the group of spatial rotations in three dimensions has no even-dimensional irreducible representations. Thus, the orbit of a particle with angular momentum j has $2j + 1$ different directions.[33] There is no known physical explanation for this; it simply follows from our Maximality Principle and the properties of the quantum mechanical formalism, neither of which has any known nonmathematical formulation.

A knowledgeable reader may here ask: what about the so-called "spin" of the electron? It has only two directions: "up" and "down." And two is an even number! The answer leads us to the most profound mystery so far: there is no contradiction here. The spin of the electron is not a spatial phenomenon, in the sense that the "spinning" electron does not consist of orbiting parts. If it did, it would be subject to the same rules as an orbiting particle.[34] The formalism dictates: any angular momentum of a particle which has an even number of directions cannot arise from spatial rotations. And this is exactly how the electron behaves.

---

[32] This space is not exactly a subspace, but a quotient space, of the entire Hilbert space.

[33] For a more technical treatment see Appendix A below.

[34] Of course the electron both orbits and "spins." Thus it has two types of angular momentum, each following its own rule.

The Hilbert or vector space formalism allows the physicist to state and prove conservation laws that have no meaning in classical mechanics, again "reading off" facts from the formalism. Consider the concept of *parity*. Suppose we represent a system by a unit vector as usual. Then the system is said to have a parity if the *mirror image* of the system has a unit vector which either stays the same or is multiplied by -1. If it stays the same, the system has positive parity; if multiplied by -1, it has negative parity. Although the concept of right-left symmetry is certainly ancient, and known to physics (e.g., in the study of crystals), the concept of parity has no analogue in classical mechanics, because (a) the unit vector in quantum mechanics is not a spatial vector; and (b) there is no physical difference between a state vector and its negative.

Now, using group theory, we can prove the following theorem, which can neither be stated nor proved in classical mechanics. If a hydrogen atom (or any other symmetrical system) has angular momentum j units, then: if j is an even number, the parity of the atom is positive; if an odd number, negative. (There is no known physical basis for this—it just follows mathematically from the quantum mechanical formalism.) Furthermore, the parity of an atom is usually[35] reversed, when it emits or absorbs a low energy[36] photon.[37] Therefore, when a hydrogen atom absorbs or emits a photon, its orbital angular momentum must usually change as well, a purely mathematical dictate, but one that has quite observable spectroscopic consequences, since it suppresses certain spectral lines. Accounting for the so-called "missing lines" of the hydrogen spectrum was one of the major research programs in quantum mechanics prior to 1926 (we shall return to the missing lines later). Here I have accounted for some of them using right-left symmetry or parity; conservation of angular momentum explains others.[38]

---

[35] Technically, the parity reversal is a "first-order" effect in the sense of perturbation theory.

[36] One whose wave length is significantly longer than the dimensions of the atom.

[37] See Sternberg 1994, secs. 3.7, 3.10, 4.5.

[38] Bohr's approach to the "missing lines" did not use symmetries, but relied on what he called the "correspondence principle." This principle does not lack "magic" of its own; for an account of Bohr's reasoning, see Darrigol 1992, 102–49.

Let us now recapitulate: beginning with the concept of a Hilbert space, a certain kind of (usually infinite-dimensional) vector space, and the formal requirement that a unit vector on the space represents[39] all possible information about a system, an astounding amount of information can be gleaned. First, the space cannot be a real vector space; the usual formalism is, therefore, based on a complex Hilbert space.[40] With this formalism, the Heisenberg Uncertainty Principle follows directly. So does the quantization of angular momentum, including the so-called "space quantization." So does the prediction that "electronic spin" cannot be due to a spatial rotation. And so do the selection rules for the spectrum of hydrogen, based on the "nonphysical" concept of parity.

The role of Hilbert spaces in quantum mechanics, then, is more profound than the descriptive role of a single concept. An entire *formalism*—the Hilbert space formalism—is matched with nature. Information about nature is being "read off" the details of the formalism. (Imagine reading off details about elementary particles from the rules of chess—castling, en passant—à la Lewis Carroll in *Through the Looking Glass*.) No physicist today understands why this is possible, though there are those who are making valiant efforts. Thus, the descriptive applicability of the Hilbert space formalism, which follows from the maximality principle, remains a mystery. To quote Feynman again:

> I find it quite amazing that it is possible to predict what will happen by mathematics, which is simply following rules which really have nothing to do with the original thing. (Feynman 1967, 171)

* * *

To eliminate the mystery of a *particular* mathematical concept describing a *particular* phenomenon, we match the concept to a nonmathematical property, as before with linearity.

[39] The term "represents" is meant in the technical sense of group representations: the unit vector, by its transformations, irreducibly represents all the symmetry groups of the physical system.

[40] The use of complex numbers is not made mathematically necessary by the arguments given here. Another possibility would be a Hilbert space with *quaternionic* coordinates. And there are physicists who have come around to the view that such a space has great potential.

Here is an example of a solved mystery. At the beginning of the sixties, both Yuval Ne'eman and Murray Gell-Mann predicted the existence of the so-called omega minus particle, using an abstract classification scheme known as SU(3), a scheme which we will discuss in detail later. This mysterious scheme turned out to be the key to classifying the strongly interacting particles—the hadrons. Later, quark theory explained the SU(3) classification by building each hadron out of the right quarks, just as the atomic theory explained the table of the elements by constructing them.

This explanation removes *one* mystery, but leaves another. Mathematicians, not physicists, developed the SU(3) concept, for reasons unconnected to particle physics. They were attempting to classify continuous groups, for their own sake.

Because the SU(3) story is not isolated, there are physicists who maintain that mathematical concepts *as a group*, considering their origin, are appropriate in physics far beyond expectation. This is a separate question from those we have been discussing, and, I believe, the most profound. It concerns the applicability of mathematics as such, not of this or that concept. It is a therefore an *epistemological* question, about the relation between Mind and the Cosmos. It is the question raised by Eugene Wigner about the "unreasonable effectiveness of mathematics in the natural sciences" (Wigner 1967).

Wigner's flawed presentation, however, hinders philosophers from giving Wigner his due. Wigner cites a number of isolated examples of mathematical concepts (e.g., the inverse square) whose effectiveness in physics is quite unexpected, given the source of the concepts in (what he claims is) aesthetics.[41] He then, in effect, offers the following invalid syllogism:

Argument A

(1) Concepts $C_1, C_2, \ldots, C_n$ are unreasonably effective.
(2) Concepts $C_1, C_2, \ldots, C_n$ are mathematical.
(3) Hence, mathematical concepts are unreasonably effective.

---

[41] Wigner's modesty prevents him from giving the most striking examples of "unreasonable effectiveness"—namely, the ones he himself discovered in applying group theory to atomic physics.

We can deduce only that *some* mathematical concepts are unreasonably effective. Further, even if the concepts are "unreasonably effective," is their effectiveness related to their being mathematical?

There is another argument Wigner is not always careful to distinguish from the first one:

Argument B

(1) Mathematical concepts arise from the aesthetic impulse in humans.

(2) It is unreasonable to expect that what arises from the aesthetic impulse in humans should be significantly effective in physics.

(3) Nevertheless, a *significant* number of these concepts are *significantly* effective in physics.

(4) Hence, mathematical concepts are unreasonably effective in physics.

Argument B does highlight the *mathematical* character of the phenomenon. But it invites the retort: what is so significant about the number of mathematical concepts that have proved effective in physics? What about all the failed attempts to apply mathematics to nature? Are not, in fact, most such attempts doomed to failure? If Wigner replies that even a single success in applying a mathematical concept is significant, he is thrown back to Argument A.

Wigner could counter that his thesis applies to the set of mathematical concepts, not to the set of attempts to apply mathematical concepts. Of the mathematical concepts, it can be said that a *significant* number of them proved *significantly* effective: They permit remarkably accurate empirical predictions, and the accuracy of these predictions tends to increase over time—with the increasing accuracy of our measuring instruments.[42] (He could also point to a sort of converse: almost every phenomenon identified before Newton—electricity, magnetism, gravity, light, the motion of fluids, etc., etc.— turned out to be describable by a mathematical law.) This formulation, though, is susceptible to challenges from skeptics who feign not to know what a "significant" number is, or when effectiveness is "significant."

---

[42] Wigner makes a point like this about Newton's law of gravitation.

Rather than rebuffing these challenges, I proceed to develop a version of Wigner's thesis to which they are irrelevant. Like Wigner's Argument B, my argument speaks to mathematical concepts in general. Unlike Wigner, however, I shall explore the peculiar role of mathematics in scientific *discovery*.[43]

First, though, it is well to take stock. I have discussed the semantic and the descriptive sense of applicability, and four associated philosophical problems:

- How can the "pure" and "mixed" contexts of arithmetic (or other mathematical theories) be understood semantically so that arguments containing both contexts can be valid? Frege solved this problem.
- How can the abstract entities of mathematics relate to the world of physics? Frege's answer was: they do not; they are related to the laws of the world, not to the world itself.
- Why are the *specific* concepts and even formalisms of mathematics *useful* in describing empirical reality? The problem must be solved piecemeal for each concept.
- Wigner's epistemological problem for mathematics as a whole: how does the mathematician—closer to the artist than to the explorer—by turning away from nature, arrive at its most appropriate descriptions?

---

[43] An early version of the argument is Steiner 1989.

3

Mathematics, Analogies, and Discovery in Physics

The obvious way of discovering laws by mathematics is to deduce them mathematically from old laws.[1] I will not add to Frege's treatment of the deductive role of mathematics.[2]

We thus ponder the *nondeductive* role of mathematics in discovering the laws of nature. This role was thrust upon mathematics by circumstances: by the end of the nineteenth century, physicists began to suspect that the alien laws of the atom could not be deduced from those governing macroscopic bodies. Nor, of course, could they be determined by direct observation. Atomic physics seemed reduced to blind guessing, with an uncertain future.

Perhaps nobody has thought harder about scientific discovery than Charles Peirce.[3] Unlike the classical empiricists, Peirce distinguished sharply between confirmation ("induction") and discovery (which he called "abduction"). Unlike the contemporary philosopher, Gilbert Harman,[4] Peirce maintained that a hypothesis, dis-

---

[1] I am using the term "laws" the way physicists do. One important feature of this use is that laws that have been refuted may still be called "laws," if they continue to describe accurately a certain class of phenomena. Thus, Newton's laws are still called that today, though they fail to describe the tiny and the rapid. I mention this only because Carl Hempel and other philosophers of science tend to reserve the term "law" for what is true, an insistence which entails that we can never be certain in science whether a hypothesis is a law or not.

[2] See above, Chapter 1. I will not discuss, in this book, whether the effectiveness of logical deduction itself needs philosophical explanation.

[3] All quotations are from Peirce 1958.

[4] See Harman 1989.

covered by "inference to the best explanation," is nothing but conjecture and requires independent confirmation.

On the other hand, Peirce noted, abduction (guessing) would be futile if the human race had not an inborn talent for hitting on the truth:

> 7.678. But just so when we experience a long series of systematically connected phenomena, suddenly the idea of the mode of connection, of the system, springs up in our minds, is forced upon us, and there is no warrant for it and no apparent explanation of how we were led so to view it. You may say that we put this and that together; but what brought those ideas out of the depths of consciousness? On this idea, which springs out upon experience of part of the system, we immediately build expectations of what is to come and assume the attitude of watching for them.

> 7.679. It is in this way that science is built up; and science would be impossible if man did not possess a tendency to conjecture rightly.

Peirce argued that the success of science to date could not be explained by chance:

> 7.680. It is idle to say that the doctrine of chances would account for man's ultimately guessing right. For if there were only a limited number n of hypotheses that man could form, so that $1/n$ would be the chance of the first hypothesis being right, still it would be a remarkable fact that man only could form n hypotheses, including in the number the hypothesis that future experimentation would confirm. Why should man's n hypotheses include the right one? The doctrine of chances could never account for that until it was in possession of statistics of the hypotheses that are inconceivable by man. But even that is not the real state of things. It is hard to say how many hypotheses a physicist could conceive to account for a phenomenon in his laboratory. He might suppose that the conjunctions of the planets had something to do with it, or some relation between the phases of variability of the stars in α Centauri or the fact of the Dowager empress having blown her nose 1 day 2 hours 34 minutes and 56 seconds after an inhabitant of Mars had died.

Peirce, therefore, looked for the explanation of this pre-established harmony between the connection of thoughts and the connection of events. Understanding the harmony would, he hoped, actually increase it.

7.506. . . . What led me into these metaphysical speculations [about the correspondence between physical and psychical laws], to which I had not before been inclined . . . was my asking myself, how are we ever going to find out anything more than we now [know] about molecules and atoms? . . .

7.507. As a first step toward the solution of that question, I began by asking myself what were the means by which we had attained so much knowledge of molecules and ether as we already had attained. I cannot here go through the analysis . . . But that knowledge has been based on the assumption that the molecules and ether are like large masses of ordinary matter. Evidently, however, that similarity has its limits. We already have positive proof that there are also wide dissimilarities; and furthermore it seems clear that nearly all that method could teach has been already learned.

7.508. We now seem launched upon a boundless ocean of possibilities. We have speculations put forth by the greatest masters of physical theorizing of which we can only say that the mere testing of any one of them would occupy a large company of able mathematicians for their whole lives; and that no one such theory seems to have an antecedent probability of being true that exceeds say one chance in a million. When we theorized about molar dynamics we were guided by our instincts. Those instincts had some tendency to be true; because they had been formed under the influence of the very laws that we were investigating. But as we penetrate further and further from the surface of nature, instinct ceases to give any decided answers; and if it did there would no longer be any reason to suppose its answers approximated to the truth.

In other words, although natural selection could possibly explain our successes in discovering science to date (since it allegedly predicts a kind of harmony between brain and world), it also predicts our inability to discover those laws (such as the laws of the atom) which had nothing to do with the evolutionary process, or survival, itself.

It is true, of course, that—if reductionism is correct[5]—the laws of the atom govern the brain's speculations about nature, as they gov-

[5] And it is not clear that reductionism *is* correct. For example, most physicists believe that chemistry is reducible to Schroedinger's equation plus the appropriate initial conditions governing the individual atoms; yet there is surprisingly little backing for this belief, owing to the enormous difficulty of carrying out the mathematics involved (cf. Primas 1983).

ern every aspect of the body. Does this make it more likely that the brain could discover the laws of the atom, than if Cartesian dualism were true? Of course not: the laws of thought, even if the severest form of materialist reductionism is true, follow from the laws of the atom *together with* an innumerable number of "initial conditions" which, from the point of view of physics, are accidental.[6] Probably, infinitely many "laws of thought" are thus consistent with the atomic laws. In other words, the "program" of the brain is not determined by those laws, any more than the program of a computer is determined by the electromagnetic laws of its hardware.[7]

Similarly, the laws of the subatomic world do not by themselves imply the laws of the macroscopic world (the laws which make a difference, for example, to survival and reproductive success), but only by way of countless initial conditions. To put the matter even more strongly: the macroscopic world which affects natural selection is consistent with innumerable possible subatomic stories.[8] This is particularly so because the kind of evidence which decides the subatomic issues is cooked up in laboratory experience using extremely sophisticated machinery. No events occurring naturally in our human environment, at the time when the human brain could have been evolving, would have decided, for example, what the symmetries of the nuclear particles are. Indeed, most of the symmetries of these particles are not exhibited at all in observable nature.

[6] For example: that the orbits of the planets lie, more or less, in a single plane could not be predicted from the law of gravity. Newton himself held that God had to be invoked to explain this. Even where the symmetries of bodies were those of the underlying laws—as is the case with the right-left symmetry of the human body—this coincidence would have to be regarded as a miracle.

It is interesting that Newton's conception of the miraculous order of nature is different from the medieval one. Newton marvels at the mathematical order, not of the laws of nature, but of the initial conditions.

[7] In what follows, I will grant Peirce, for the sake of argument, that he is right about the laws of the macroscopic environment: namely, that we have a disposition to discover them because of their conspicuous role in human evolution. I confess, though, that I have difficulty seeing why, for example, because the laws of thermodynamics play a role in the evolution of the brain, the brain has an enhanced capacity to discover these laws. I will leave the matter open.

[8] This claim might sound like a standard anti-realist line, but it isn't. The antirealist line is that observations cannot decide theoretical issues; my claim is that events occurring naturally in the human environment cannot decide theoretical issues in areas like particle physics.

Thus, if we are to acquire an ability to guess correctly at the laws of atomic physics, we must go beyond natural selection. Accordingly, Peirce adopted a hybrid metaphysics reminiscent both of Spinoza and of Hegel, but which outstrips both of them.

Here's the theory: evolution encompasses not just the biological sphere, but the entire universe. The laws of this evolution operate everywhere, and govern also the evolution of mind.[9] What is more, the operation of both mind and the world reflect their (common) evolutionary history. Thus, by studying the operation of mind, one can discover the laws of the world by applying these operations to existing physical theories. Here is a sample of Peirce's thinking:

> 7.515. But if the laws of nature are results of evolution, this evolution must proceed according to some principle; and this principle will itself be of the nature of a law. But it must be such a law that it can evolve or develop itself. . . . Then the problem was to imagine any kind of a law or tendency which would thus have a tendency to strengthen itself. Evidently it must be a tendency toward generalization,—a generalizing tendency. But any fundamental universal tendency ought to manifest itself in nature. Where shall we look for it? We could not expect to find it in such phenomena as gravitation where the evolution has so nearly approached its ultimate limit, that nothing even simulating irregularity can be found in it. But we must search for this generalizing tendency rather in such departments of nature where we find plasticity and evolution still at work. The most plastic of all things is the human mind, and next after that comes the organic world....Now the generalizing tendency is the great law of mind, the law of association, the law of habit taking. We also find in all active protoplasm a tendency to take habits. Hence I was led to the hypothesis that the laws of the universe have been formed under a universal tendency of all things toward generalization and habit-taking.

Peirce's ideas are brilliant, if wacky. Yet his problem was real, and it is well to study its historical resolution. How did physicists discover

---

[9] These universal super-laws were, to Peirce's thinking, the key to the formal mathematical analogies we see between laws—such as the inverse square laws in gravity and electricity—analogies that demand explanation (7.509–7.511). But Peirce looked to these super-laws also to explain, not only the mathematical form of laws, but even the specific values of the constants (like the gravitational constant) appearing in them.

successful theories concerning objects remote from perception and from processes which could have participated in Natural Selection?[10]

My answer: by analogy. Having no choice, physicists attempted to frame theories "similar" to the ones they were supposed to replace. But the writings of Nelson Goodman and Ludwig Wittgenstein[11] prompt the question: in what respect "similar"?[12] Any objects are "similar" in some respects and "dissimilar" in others. An analogy, therefore, presupposes a taxonomy—a scheme of classifying. The answer, "by analogy," is so far no answer, unless the ground of the analogy be set forth.[13]

[10] Curiously, contemporary physics—without realizing it—did adopt, with success, a version of Peirce's procedure. *One* of the avenues to progress in physics has been the following "rule" which I shall call the Peirce/Steiner rule (I'm only kidding):

Suppose a theory T utilizes mathematical concept C. To develop T, we look to the history of mathematics, where we find that concept C been generalized to C*. Then we generalize theory T by replacing concept C by C* in theory T.

That is, we suppose, and exploit, a kind of recapitulation of the evolution of mathematics in the evolution of mathematical physics!

After writing the above, I came across the following declaration by a contemporary physicist: "I believe that the following is a true and somewhat mysterious fact: deeper physics is described by deeper mathematics" (Zee 1990). This is a generalization of "my" rule. Whether or not either rule is "mysterious" is what I will now consider.

[11] Analogies between the work of these two, apparently dissimilar, philosophers were noted in Kripke 1982.

[12] Goodman's "New Riddle of Induction" (Goodman 1983, chs. 3 and 4) deals primarily with the question, "What is the criterion for things to be truly alike?" In the *Philosophical Investigations*, Wittgenstein argues that "The use of the word 'rule' and the use of the word 'same' are interwoven" (Wittgenstein 1968, § 225), and that the perplexities surrounding the concept of rule-following, dog also that of similarity.

[13] Wittgenstein, it is true, does say "When I obey a rule, I do not choose. I obey the rule *blindly*" (Wittgenstein 1968, § 219). And since "The use of the word 'rule' and the use of the word 'same' are interwoven" (§ 225), it follows that making comparisons, like following rules, is not done by deliberation. But it would be absurd to saddle Wittgenstein with the view that we *never* think when following a rule; or, indeed, that we never err when following a rule. That erring is possible, according to Wittgenstein, is part of the very grammar of following a rule (§ 201); and, in following a rule, "a doubt was possible in certain circumstances" (§ 213). After all, we do have a concept of superstition; we often condemn people who generalize blindly.

Wittgenstein's point is simply that "there is a way of grasping a rule which is *not* an *interpretation*" (§201), and that the entire practice of rule-following could not exist unless some rules are grasped this way (i.e., followed "blindly"). Generally, we condemn people who follow rules blindly when there are higher-order rules that dictate how lower-order rules are to be followed. These kinds of rules exist in sophisticated practices such as court cases and in scientific practice. Wittgenstein's message is that "Explanations come to an end somewhere" (§ 1). Thus it is open to me, for example,

Besides, as Peirce pointed out, reasoning by *physical* analogy had already been discredited in atomic theory. The whole trouble was that the laws (if any) of the atom (if any) were proving *not* to be analogous to those of bodies. The answer can only be, that (for lack of anything better) scientists began relying on nonphysical analogies.

I shall portray two kinds of analogy, or taxonomy, that recur in the reasoning of the great inquirers: the one I call "Pythagorean"; the other, "formalist." About these, I make two claims: first, that they are deeply "antinaturalist"; second, that without them, contemporary physics would not exist. We now have three terms to define: "Pythagorean," "formalist," and "naturalist."

By a "Pythagorean" analogy or taxonomy at time *t*, I mean a mathematical analogy between physical laws (or other descriptions) not paraphrasable at *t* into nonmathematical language.[14] Previously we had examined physically based mathematical *concepts* (linearity) and also concepts (analyticity, Hilbert space) which, presently, are not so based.[15]

By a "formalist" analogy or taxonomy, I mean one based on the syntax or even orthography of the *language* or *notation* of physical theories, rather than what (if anything) it expresses.[16] Because any notation has, itself, a mathematical structure, formalist analogies are also Pythagorean.[17] I single out formalist analogies because, from the "naturalist" standpoint, formalist analogies are (or should be) particularly repugnant.

---

to criticize the physicists, in making blind analogies, in terms of higher-order rules which they themselves accept—as I point out now in the text.

Note, once again, the powerful similarity between the thoughts of Wittgenstein and Goodman.

[14] Though there are enormous difficulties in analyzing the term "mathematics," for reasons sketched below, there is sufficient agreement on *particular* descriptions—as to whether they are mathematical or not—to give nontrivial content to this definition. Nor does the existence of borderline cases vitiate the distinction—if it did, most distinctions, including that between good and evil, would collapse—as Nelson Goodman pointed out a long time ago.

[15] Since we are dealing with epistemology, the relativity to knowledge is expected.

[16] I am thinking here of cases (such as quantum electrodynamics) where the mathematical formalism of a theory is suspected of inconsistency so that, strictly and semantically, it expresses nothing.

[17] I am grateful to Shaughan Lavine for pointing this out to me.

"Naturalism," more an ideology than a thesis, has been variously defined.[18] Perhaps the most influential, if not the most useful, definition of naturalism is that of Quine: naturalism demands that philosophy be part of, or continuous with, natural science. (It makes no demands on science.) I see naturalism, rather, in opposition to *anthropocentrism*—the teaching that the human race is in some way privileged, central to the scheme of things. Accordingly, I will *define* naturalism to be "opposition to anthropocentrism." For example, the naturalist contends that the universe is indifferent to the goals and the values of humanity, the illusion to the contrary being fostered by natural selection or by hubris.

This definition makes naturalism—like determinism or mechanism—a regulative ideal for science as well as philosophy. For example, Steven Jay Gould regularly decries certain aspects of current evolutionary doctrine as anthropocentric—and his criticism, I would say, is naturalistic. I also think that this definition can do much of the work of the other definitions of philosophical naturalism.

Nevertheless, I have no desire to lay claim to the *word* naturalism.[19] My topic is anthropocentrism, and my goal in this book is to show in what way scientists have—quite recently and quite successfully—adopted an anthropocentric point of view in applying mathematics.

First, let us distinguish between *overt* and *covert* anthropocentrism.

Overt anthropocentrism takes the form of *theories* which preach the centrality of the human race, either explicitly or implicitly. Creationism is an example of *explicit* anthropocentrism. A theory is *implicitly* anthropocentric if it implies, via unexpressed assumptions, the privileged nature of the human race. Geocentrism, though it says nothing about the human race explicitly, is often thought to be implicitly anthropocentric.[20] Because the anthropocentrism here,

---

[18] The earliest use of the term I have discovered is in Kant's *Prolegomena* (Kant 1950, 111): "The cosmological Ideas, by the obvious insufficiency of all possible knowledge of nature to satisfy reason in its legitimate inquiry, serve in the same manner to keep us from naturalism, which asserts nature to be sufficient for itself."

[19] If there were a decent word in English for "anti-anthropocentrism," I would gladly drop the term "naturalism."

[20] Actually, although modern writers (most blatantly, Freud) look back at the geo-

though implicit, is still theory-based, I choose to call it overt anthropocentrism as well.

Covert anthropocentrism is *behavior* (not necessarily that of making assertions) which presupposes some anthropocentric doctrine. That is, behavior by an agent A is covertly anthropocentric if it is irrational if A has no anthropocentric beliefs. Our main example of covert anthropocentrism in what follows will be: classifying phenomena by reference to human peculiarities. No taxonomy is neutral—employing a taxonomy seriously can be rational only if you accept its underlying rationale.

It is easy to document the revolt against (even) covert anthropocentrism in modern science. The Galilean–Newtonian revolution, for example, not only attacked the overt geocentrism of medieval science, it undermined the very classification of events into "heavenly" and "terrestrial."[21] And no contemporary physicist would classify elementary particles as male and female, although the ancient Pythagorean systems did so classify numbers,[22] and therefore phenomena in general.[23]

---

centric universe as anthropocentric, even narcissistic, the medievals themselves did not necessarily see geocentrism as noble. On the contrary: though a few writers (Saadyah, for example) *did* see geocentrism as evidence that the universe was created for man, many others (like Maimonides) saw geocentrism as a doctrine of humiliation. In many cosmologies the center of the universe is the worst place to be, a cosmic dungeon or even dungheap (for a recent treatment of this idea, with citations from Islamic and Christian, as well as Jewish, philosophy, see Brague 1990).

Nevertheless, I believe that the very idea of a special place for the human race—even a bad place—is anthropocentric, in the sense that Somebody cares enough about us to *put* us in a special place. (As the Yiddish saying goes, "You're not so big—don't make yourself so small.") The Copernican revolution leaves homo sapiens without *any* place; the universe couldn't care less where we are. In any case, this book is not about anthropocentric *doctrines*, as the reader will soon see; it is about *anthropocentric schemes of classification*. Whether it is good or bad to be at the astronomical center, the mistaken idea of classifying events as either "terrestrial" or "celestial" does arise from the geocentric universe. And the geocentric universe arises from the viewing of events from the point of view of the earth, which nobody would do if we didn't live here. Thus, geocentrism is, in fact, anthropocentric, even if not noble.

[21] Indeed, once the heavenly/earthly classification was eliminated by implication in Newton's laws of motion, which made no differentiation between forces operating on terrestrial or celestial bodies, the refutation of *overt* geocentrism became trivial: no body could exert the kind of forces which would be needed, on Newton's theory, to keep the entire universe circling around it every 24 hours.

[22] Odd numbers were male; even, female. Cf. also p. 6 above.

[23] Yuval Ne'eman writes that, in attempting to classify the hadrons, he had

Covert anthropocentrism can lead—today—to the dismissal of a hypothesis as unworthy of attention. In fact, covert anthropocentrism is often equated with superstition. For example, the following "hypothesis" was "confirmed" seven times before its refutation by Ronald Reagan:

(P) Beginning with 1840, the President of the United States elected in a year divisible by 20 dies when in office.

Should a rational politician have refrained from running for President of the United States in 1980 for this reason? Such a fear is superstitious, we feel. Philosophers of science put things as follows: hypothesis (P) is *not* confirmed, because it is unconfirmable. In Nelson Goodman's terminology (Goodman 1983), hypothesis (P) is "unprojectible." The seven "successes" are a fluke.

Why is (P) unprojectible? We do not regard "President of the United States," a category invented by (one) human society for its own purposes, as relevant to mortality.[24] Similarly, whether the year (of the Gregorian calendar) is divisible by 20 (or even 2,000) or not makes no difference to nature. Nor is there anything special about "round" numbers like 20, which are "round" only in our parochial decimal system, useful to us because we have ten fingers. Finally, since the category "dies in office" includes death by "natural" causes and death by assassination (and also refers to a naturalistically irrelevant "term of office"), we don't expect any laws applicable to this artificial class.

In sum: hypothesis (P) is not even a candidate for confirmation, because it is *covertly anthropocentric*: the concepts it applies presuppose that the human race enjoys a special status.[25] The act of

---

"hoped" that the Star of David would play a role. As far as I know, this is the first attempt to apply the Jewish–gentile distinction in physics. (Ideologists have attempted to apply national distinctions in classifying *types of physical theories*, another issue.)

[24] I have heard that the Mormon religion regards the Constitution of the United States as divinely inspired; for a Mormon, the question of projectibility of (P) would look completely different.

[25] And here, "special" does not necessarily mean "preferred."

The knowledgeable reader will wonder why, if I mention Nelson Goodman, I leave out his concept of "entrenchment" (cf. Goodman 1983, ch. IV, for this concept). The reason is simple: "entrenchment," in Goodman's theory, is significant at the most basic level of projection. As Goodman himself points out, however, background

projecting hypothesis (P), in other words, makes no sense except on anthropocentric principles. For on naturalist principles, (P) is unprojectible—and one cannot make rational predictions based upon unprojectible hypotheses, even if the hypotheses have been consistent with the data up till now.

Covert anthropocentrism, then, makes the same kind of statement as overt anthropocentrism. Those who loathe anthropocentrism ought to loathe covert anthropocentrism all the more for being insidious.

There is, indeed, a kind of anthropocentrism—I'll call it "play it safe" anthropocentrism—which pretends to avoid statements concerning the status of the human race in the Great Chain of Being. "Play it safe" anthropocentrism advocates, modestly, that science cannot confirm any hypothesis about the unobservable. In Van Fraassen's words, all we can reasonably hope to do is to confirm by evidence that scientific hypotheses are "empirically adequate," meaning roughly that they are consistent with all possible human observations.[26] Van Fraassen admits that this term is anthropocentric, but sees no problem: human science can only reflect human limitations. In any case, he argues, the realist alternative sees science as aiming at truth, and empirical adequacy is just a weakening of that goal. Hence empirical adequacy is a safer bet than truth,[27] and therefore the more rational conclusion—always.

I regard "play it safe" anthropocentrism as no different than any other kind. A philosopher who objects to projecting anthropocentric hypotheses should also object to *limiting* or *weakening* hypotheses by restricting them to anthropocentric categories (such as empirical

---

hypotheses—"overhypotheses," as he calls them—will play the dominant role in mature sciences. Goodman emphasizes "entrenchment" only because it is overlooked by philosophers who realize the importance of background information in projectibility judgments, but overlook the question of the projectibility of the background information itself! Arguing by regress, Goodman arrives at "entrenchment," a property that is independent of background hypotheses. (For a very different treatment of the role of background information in science, see Levi 1980.)

[26] See Van Fraassen 1980.

[27] Literally, of course, the hypothesis of truth is just as safe as empirical adequacy, because any detectable refutation of a hypothesis is also a detectable refutation that the hypothesis is empirically adequate. From that point of view, why not go for truth?

adequacy). The content of the prediction may now be weaker; but the prediction itself (the speech act) makes a *covert* anthropocentric statement.

For example, the same theories that tell us that we cannot see x-rays tell us also that what we do see (light) is a random sample of radiation; i.e., that there is nothing special about light, and that there is nothing wrong per se with projecting the properties of light on unobservable radiation. Deliberately to limit our projections concerning radiation to light, then, is to imply,[28] covertly, theses about the centrality of the human race—theses which are far more consequential than the "realist" conclusions (about the relevance of observation to the unobservable) that Van Fraassen wanted to avoid. From this point of view, the assumption of empirical adequacy (as distinguished from truth) is not "playing it safe" at all. It is adopting an anthropocentric world view, though covertly.[29]

Finally, *projecting* (or failing to project) a hypothesis is only one activity one performs with a taxonomy. Guessing by analogy is another. Nevertheless, where the analogy is improper, as anthropocentric analogies are *for the naturalist*, guessing by analogy can be just as irrational as projecting an unprojectible hypothesis. I take it that using a dowsing stick to pick the place to drill for oil is superstitious, even though no hypotheses are projected.

* * *

There is no question that the *modern* scientific revolution involved a revolt against anthropocentrism. Nevertheless, I now argue, *recent* physics—from about 1850—has retreated from naturalism. The truly great discoveries in contemporary physics were made possible only

---

[28] The implication referred to here is sometimes called "implicature" in the philosophy literature. For example, when asked for my evaluation of a student and I praise his handwriting, the implicature is that my evaluation is low. Implicature is a relation between an act (including a speech act) and a proposition, namely that the act doesn't make sense unless the proposition is believed.

[29] To prevent misunderstanding, I emphasize that I am not arguing against anthropocentrism—quite the contrary, this book will end up supporting it. What I object to is Van Fraassen's idea that restricting science to the empirically adequate has no philosophical implications concerning the status of the human race.

by abandoning—often covertly and even unconsciously[30]—the naturalistic point of view. This astonishing thesis follows from the following two premises:

I. Both the Pythagorean and formalist systems are anthropocentric; nevertheless,

II. Both Pythagorean and formalist analogies played a crucial role in the fundamental physical discoveries in this century.

Leaving the discussion of II for the next chapter, here's my argument for I.

Historically, Pythagoreanism was not viewed by its proponents as anthropocentric. I contend that no *modern* version of Pythagoreanism avoids this charge: mathematics has changed considerably since Pythagoras. Pythagoreanism began as a doctrine concerning numbers; but the term "Pythagoreanism" in this book does not reflect bias toward number theory. To the contrary, number theory has played little role in physics; the characteristic "Pythagorean" concepts—prime numbers, perfect numbers—have surfaced neither in physics nor in physical discovery.[31]

Also, historical Pythagoreanism was primarily metaphysics; I accent its epistemology. Thus, I shall not discuss whether the world "is" numbers—or, for that matter, an irreducible representation of some group. Pythagoreanism for me is the teaching that the ultimate "natural kinds" in science are those of pure mathematics.[32]

The discovery that musical intervals map into the natural numbers was an early triumph of a creed not usually associated with empirical inquiry. In late antiquity, a synthesis of Pythagoreanism with

[30] Behavior can often go contrary to, be irrational in light of, one's professed beliefs (no atheists in foxholes). Since practicing scientists often advocate naturalism, at least in my sense (Hawking 1988 is an example), their behavior in making discoveries is unintelligible, given their professed views. We focus our attention here on what scientists do, not what they say.

[31] Recently, however, work has been done on the problem: what would a bound quantum system be like, if its discrete energy levels were given by the zeros of the Riemann zeta function. (The Riemann zeta function is well known to carry information about the prime numbers.)

[32] Or, weaker: that *some* of the ultimate "natural kinds" are those of mathematics. This could be called "weak" Pythagoreanism, with "strong" Pythagorean the proposal mentioned in the text.

Aristotelian science (the "four causes") emerged. I quote from Iamblichus, of the third century C.E.:

> There is an efficient cause in physical numbers: one may see this in the generative numbers shown in animal generation. And the principle of movement according to difference and inequality in numbers shows an efficient cause. But this is especially manifest in the rotations and the revolutions of the heavens. And the stars' configurations in relation to each other, their periodic revolutions, all of their shapes, their powers, are contained in the principles of numbers. And the moon's phases, the order of the spheres, the distances between them, the centres of the circles which carry them, numbers contain them all. Indeed the measures of numbers determine health; crises in sickness are completed according to determinate numbers; deaths come thus also, nature having fulfilled the appropriate measures of change. Hence number is generative of animal life. For since animals are made up of soul and body, the Pythagoreans say soul and body are not produced from the same number, but soul from cubic number, body from the bomiskos. For, they say, <soul's> being is from equal times the equal times, coming to be in equality, whereas body is a bomiskos, produced from unequal times the unequal times. For our body has unequal dimensions: its length is greatest, its depth least, its breadth intermediate. Thus soul, as they say, being a cube from the number 6 (which is perfect), comes to be equal an equal times the equal as in the cube 216, for this is 6 by 6 by 6. But body, being from unequal sides an unequal times the unequal an unequal times, is neither dokis nor plinthis but a bomiskos, having for sides 5, 6, 7: for 5 by 6 is 30, and 7 by 30 is 210. Thus seven-month births occur in 210 days, having a complete body. If then the soul alone were generated, it would be born in 216 days, a perfect cube being completed with its coming. But since the animal is made of soul and body, 210 days are appropriate to the completion of the body: the generation of the body dominates in the animal. Thus soul desires equality, the body relates to anomaly and inequality.[33]

What is consequential here is not the reference to numbers as "objects," but the number-theoretic taxonomy: cubic number, bomiskos, perfect number, etc. The soul is associated with the cube of a perfect number, 6. The body, on the other hand, is associated with the number $5 \times 6 \times 7$, a bomiskos in which the perfect number

---

[33] Translated in O'Meara 1989, Appendix.

is in the middle. Iamblichus has no nonmathematical explanation of the alleged numerical relationship between body and soul.

Not every mathematical classification is Pythagorean. We have seen that the classification of phenomena into *linear* and *nonlinear* can be interpreted in nonmathematical terminology. Einstein used the mathematical trait of "covariance" to express a physical idea: space has no causal properties.

Nevertheless, Pythagorean reasoning dominates twentieth-century physics, in that *Pythagorean* mathematical analogies—ones with no physical basis when made—have been indispensable in recent *discovery*. My case for this depends primarily on historical evidence; but the following logical considerations boost its *a priori* plausibility.

Consider the analogies a physicist draws. The theorist proposes, for testing, mathematical laws—themselves, analogies between the future and the past. Laws I call *first-order* analogies. In *discovering* those laws, however, one employs more abstract analogies than what the laws express. For example, physicists may look for laws with the mathematical properties of known, successful laws.[34] Such mathematical analogies are *second-order* because they are based on *properties* of laws. The physicist may resort even to a *third-order* mathematical analogy, based upon the properties of properties of descriptions. Murray Gell-Mann argued formally from the prevalence of a symmetry to that of a related, more general, one.[35]

Now the higher the order of a mathematical analogy, the more likely it is to be Pythagorean. A third-order analogy is based on the similarity between two mathematical, not material, structures. At the time the analogy is drawn (and we do not care here about any other

[34] In order to avoid getting involved in disputes about scientific realism and anti-realism, I shall not attempt to characterize too precisely what constitutes a "successful" prediction, or a "successful" mathematical description of nature.

[35] See Gell-Mann 1987. I am grateful to Yuval Ne'eman for making available to me this and many other articles on the Eightfold Way and for a number of illuminating communications (written and oral) on the subject. Though Ne'eman, independently of Gell-Mann, also discovered the Eightfold Way, his reasoning (as outlined, e.g., in Ne'eman 1987, does not seem to have involved mathematical *analogies*—and, therefore, his name does not appear in the text. I hope it is clear, finally, that an enormous amount of experimental and theoretical preparation lay behind the discovery of the omega minus; I am highlighting a specific aspect of a complicated story.

time), it is doubtful that such an analogy could have been physically grounded; otherwise, the physicist would not have had to draw it.

Now, what is anti-naturalist about Pythagorean analogies and taxonomies? This question hangs upon another, weighty one: what is mathematics? Or: what is the criterion for concepts to be "mathematical"?

From a broadly Fregean standpoint, we can distinguish three categories of mathematical concepts, depending upon whether they are (or purport to be):[36]

(1) Properties of mathematical objects (e.g., "prime number");
(2) Properties of sets or systems (e.g., "group");
(3) Second-order properties or functions (e.g., "the number of Fs," "the derivative of f").

Frege himself was interested in the differences between the categories. But we can also ask, of each category, what makes a *member* of that category a *mathematical* concept? So we get three questions, and we should not expect the answers to be the same.

For example, given (2), we can inquire after the distinction between the concept of a group and that of a game like chess. Why is the "theorem" that mate cannot be forced with a king and two knights against a king not a theorem of mathematics (and no mathematician I consulted says it is)?[37] Why are the notions of chess, like "castling," "en passant capture," "queening," etc., of no mathematical significance? The standard philosophies of mathematics—logicism,[38] formalism, or intuitionism—have no answer, since the distinction between mathematics and chess is a predilection of mathematicians, rather than a logical distinction.

[36] Discussions with Sidney Morgenbesser were invaluable in helping me formulate the following ideas.

[37] Of course, the theory of games in general is a branch of mathematics. It could well be that the theory of games, as well as other branches of mathematics, has implications for chess, in which we could speak of *applications* of these theories to specific games. It remains, however, that the "axioms" of chess are not regarded as mathematical axioms.

[38] Frege did not treat this subject: when he accused formalist Hilbert of slighting the distinction between mathematics and games, he meant only that Hilbert ignored the distinction between *proving* a theorem and *playing* a game, or between the steps of the proof and the moves of the game—not the distinction at issue here.

This attitude of mathematicians would change if new structural facts about chess emerged. Suppose that every winning position in chess had a geometrical symmetry. A "theorem" of that sort might be regarded as mathematics. Not that any mathematician expects this: chess, like many two-party games, is an abstraction of war. The provenance of chess does not augur well, by past experience, for the richness typical of mathematical concepts.[39]

Mathematics also has its roots in various human activities (measuring, counting, locomotion). But modern mathematics has traveled far from these roots. As Mac Lane puts it:

> The genesis of the more complex mathematical structures tends to take place within Mathematics itself. Here there are a variety of processes which may generate new ideas and new notions. These are: conundrums (36), completion (37), invariance (37), common structure (analogy) (37), intrinsic structure (38), generalization (38), abstraction (38), axiomatization (39), the analysis of proof (39).[40]

That modern mathematical concepts breed within mathematics (rather than empirically) is important; but what makes them *mathematical*? Most mathematicians would accept Wigner's view (Wigner 1967): modern mathematics expresses the human aesthetic sense. Concepts are selected as mathematical because they foster beautiful theorems and beautiful theories. That the processes listed by Mac Lane generate beauty is a remarkable, contingent fact about the history of mathematics. In the words of a great twentieth-century mathematician:

---

[39] Recent computer analysis of chess, working backwards, has revealed surprising results. There are positions which result in a win for one of the players in over 90 moves, none of which involves the capture of a piece or the movement of a pawn. The winning "line," moreover, is almost impossible to memorize, because there seems to be no rhyme or reason for the specific winning moves. Seemingly random moves result in a win, without anyone being able to characterize the moves as a "strategy" of one kind or the other. This is the opposite of what we find in mathematics.

[40] Mac Lane 1986, 36. The numbers in parentheses are the author's page references to that work.

Note that the "processes" listed here presuppose a system of taxonomy or classification: consider particularly "invariance" and "analogy." Thus, from our point of view, any use of Mac Lane's list in order to characterize mathematics as "objective" (nonanthropocentric) simply begs the question (I don't claim, of course that Mac Lane himself intended to do this).

The mathematician has a wide variety of fields to which he may turn, and he enjoys a very considerable freedom in what he does with them. To come to the decisive point: I think that it is correct to say that his criteria of selection, and also those of success are mainly aesthetical. . . . One expects a mathematical theory not only to describe and classify in a simple and elegant way numerous and a priori disparate cases. One also expects "elegance" in its "architectural," structural makeup. . . . These criteria are clearly those of any creative art. (Von Neumann 1956, 2062)

Of course, G.H. Hardy is notorious for the view that beauty is the essence of mathematics:

The mathematician's patterns, like the painter's or the poet's, must be *beautiful*; the ideas, like the colours or the words, must fit together in a harmonious way. Beauty is the first test: there is no permanent place in the world for ugly mathematics . . ..

It may be very hard to *define* mathematical beauty, but that is just as true of beauty of any kind—we may not know quite what we mean by a beautiful poem, but that does not prevent us from recognizing one when we read it.[41]

That the aesthetic factor in mathematics is constitutive has actually become a truism in the mathematical community. A survey in a recent issue of the *Mathematical Intelligencer* asked readers to rank mathematical theorems in order of their beauty (Wells 1988). Perhaps the question was flawed for divorcing the beauty of a theorem from the "architecture" of its proof. Still, I doubt that a periodical in any other science would conduct such a survey.

[41] Hardy 1967, 85. Hardy goes on to say that the solution to a chess problem is (or can be) mathematics, because it partakes of the same kind of beauty as mathematics, and in this he seems to differ from my view. But he immediately goes on (p. 88) to say that a chess problem is "trivial" mathematics. This is because chess problems are not *serious* in the way mathematics is. But he then adds (p. 90) that "the beauty of a mathematical theorem *depends* a great deal on its seriousness"! This almost circular reasoning brings him right back to my position.

Nor does Hardy's insistence that pure mathematics has no utility (probably because utility compromises beauty) have any bearing on its applicability. Utility and applicability, he rightly holds, are two separate things. Hardy explicitly recognizes Maxwell, Einstein, Dirac, and Eddington as "real" mathematicians. Quantum mechanics and relativity, he says, are applied mathematics, but at present are "almost as 'useless' as the theory of numbers" (131). We can, of course, today laugh (or cry) at the naivety of the last sentence.

It seems plausible, then, that the best answer to the question: "Why is chess a game; but Hilbert spaces, mathematics?" will rely on aesthetics. What *caused* mathematicians to frame the concepts they did was often their taste. This certainly accords with Pythagoreanism, which was, *inter alia*, an influential *aesthetic* doctrine, centuries after Pythagoras.[42] For me, though, the mathematical sense reduces to the aesthetic, whereas historical Pythagoreanism, limiting itself to one structure only (the natural numbers), claimed the reverse. To say that the mathematical sense reduces to the aesthetic is to deprive the aesthetic sense of the only argument for its objectivity—namely, that the aesthetic sense is based on the objectivity of mathematical form, as the Pythagoreans in fact argued. If the Pythagorean position on aesthetics *today* begs the question—if, as I hold, the term "mathematical form" (given the multitude of "mathematical forms" today) is empty without introducing the human aesthetic sense—then there is no escape from the conclusion that the human aesthetic sense is nothing but species-specific preference. Classifications like beautiful/ugly are then anthropocentric; so, finally, are the mathematical classifications.

Besides aesthetic considerations for mathematical concepts, there is another—convenience. Our power to compute, like every human power, is limited; computational aids compensate. For example, imaginary numbers were introduced by Cardano as solutions of quadratic equations. Bombelli noted (what Cardano probably also knew) that Cardano's formula for the roots of a cubic equation can make even a real root the sum of the square roots of two imaginary numbers. As Roger Penrose puts it:

> While at first it may seem that the introduction of such square roots of negative numbers was just a device—a mathematical invention designed to achieve a specific purpose—it later becomes clear that these objects are achieving far more than that for which they were originally designed. As I mentioned above, although the original purpose of introducing complex numbers was to enable square roots to be taken with impunity, by introducing such numbers we find that we get, as a bonus, the potentiality for taking any other kind of root or for solving any algebraic equation whatever. Later we find many other magical properties that these complex

[42] See Hersey 1976.

numbers possess, properties that we had no inkling about at first. These properties are just *there*. They were not put there by Cardano, nor by Bombelli, nor Wallis, nor Coates, nor Euler, nor Wessel, nor Gauss, despite the undoubted farsightedness of these, and other, great mathematicians; such magic was inherent in the very structure that they gradually uncovered. When Cardano introduced his complex numbers, he could have had no inkling of the many magical properties which were to follow—properties which go under various names, such as the Cauchy integral formula, the Riemann mapping theorem, the Lewy extension property. These, and many other remarkable facts, are properties of the very numbers, with no additional modifications whatever, that Cardano had first encountered in about 1539. (Penrose 1989, 94–5)

Penrose here supports Platonism in mathematics;[43] a concept introduced for convenience turns out to have the characteristic "magic" of mathematics. But the same happens in applications to physics: a mathematical concept introduced for convenience turns out to have "physical reality."

Take the very case of "imaginary numbers." Before the nineteenth century, these had little use in physics. Then, complex-valued exponential functions replaced trigonometric functions in describing waves, exponentials being easier to calculate with (the derivative of an exponential is an exponential). The "imaginary part" of these functions played as yet no role.

The concept of a potential (from which fields can be retrieved by taking derivatives) also began as a computational convenience. For example, a scalar potential is a scalar function $\phi(x,y,z)$, from which we can retrieve a field as the *vector* function

$$\left(\frac{\partial\phi(x,y,z)}{\partial x}, \frac{\partial\phi(x,y,z)}{\partial y}, \frac{\partial\phi(x,y,z)}{\partial z} \right)$$

—$\Delta\phi$ for short. It is obvious that it is easier to calculate with the scalar function.

---

[43] In another essay, however, I argue that the real spiritual ancestor of Penrose's doctrine is Descartes, not Plato. It was the former, not the latter, who distinguished between "true and immutable" essences and arbitrary combinations of properties. It was David Shatz who pointed out to me the relevance of Descartes.

Finally, consider the concept of a Taylor series, i.e., the expansion of a function as a power series,

$$\sum_{i=0}^{\infty} c_i x^i.^{44}$$

This mathematical idea has great use in calculation: we calculate the value of a function to any degree of desired approximation by calculating the values of the appropriate number of terms of the expansion. If the convergence of the series is very rapid, then the first few terms by themselves can give an excellent numerical approximation to the exact value of the function. Taylor series are so useful that scientists invent them when they don't even exist. For example, they pretend that constants are variables and then "expand" in powers of the constant. (This is really a case of the formalism leading the scientist, the tail wagging the dog—we shall have occasion to see the remarkable consequences of this fiction.)

Consider the fundamental charge of the electron, e. Scientists find it immensely useful to write expressions like this:

$$f(x) = \sum_{i=0}^{\infty} a_i(x)e^i$$

derived by pretending that the charge is a variable. Each of the coefficients is a function of x. Since e is known and sufficiently small, the series converges rapidly, i.e., usefully.

One such function might be, say, $P(q \rightarrow q')$—the probability that an "excited" atom in a state q will radiate a photon and drop to state q'.[45] Photons so emitted (in large numbers) are what produce spectral lines which indicate, not the state of the atom (as nineteenth-century physicists had imagined), but transitions from one state to another. One of the historic missions of quantum mechanics was to explain why certain spectral lines were "missing" from the spectrum

[44] My understanding of these matters was greatly enhanced by conversations with Shlomo Sternberg and Shmuel Elitzur. For a profound treatment of the group theoretical issues, cf. Sternberg 1994.

[45] One might ask, how can an atom, in a defined state, make a transition? Isn't it the case that defined states in quantum mechanics are stable? The reason that atoms make such transitions is that they are surrounded, like everything in the world, by a field of electromagnetic radiation. What is an exactly defined state of the atom does not count as an exactly defined state of the combined atom-plus-electromagnetic field. This combined state can then develop over time to another combined state in which the atom is then found in another of *its* defined states.

of a given atom. I would like to discuss the vital contribution of the concept of a Taylor series: the missing lines do occur, but are invisible. To calculate the probability of the transition, we start with a Taylor series (never mind for the moment where it comes from):

$$\sum_{i=1}^{\infty} a_i(q,q')e^i$$

In quantum mechanics, the terms are complex numbers, so to finish the calculation, we take the *absolute square* of the sum:

$$P(q \rightarrow q') = \left| \sum_{i=1}^{\infty} a_i(q,q')e^i \right|^2$$

Now for certain transitions $q \rightarrow q'$ the first-order term, $a_1(q,q')e$, is zero since the coefficient vanishes. Only the first-order term could produce a transition strong enough to be detectable; whence the "missing" lines.

It is convenient—and therefore customary—to use the concept of the Taylor series in classifying phenomena as "first-order," "second-order," etc. That is, suppose we have a function that we feel tells the "whole truth" about nature. If we expand it as a series, we can say what nature would be like if we ignore terms beyond the first-order, second-order, etc., terms. In our case, then, we can say: there is no "first-order amplitude" for certain atomic transitions. Spectral lines, in other words, are first-order phenomena. So-called "Raman scattering," by contrast, is a second-order phenomenon, since the first-order term of the Taylor series does not allow for such scattering. The second-order term does, and this probability contains a factor of e^2, and hence is very small.

The classification of *phenomena* (as distinct from terms) as first-order, second-order, etc., is certainly not one that derives from observation, but convenience. We could just as well have classified phenomena according to decimal powers: i.e., magnitudes from 0 to 9 belong in one class, from 10 to 99 in the next class, etc. The real distinction is that between first-order, second-order, . . . *terms*—a mathematical distinction deriving from the calculus of Taylor series. Projecting the distinction on the phenomena is anthropocentric, but scientists do it anyhow.

More generally, they like to assume that what is convenient is correct (a trait by no means restricted to scientists). They hope, for example, that their *calculations* will end in a blaze of cancellations, as a sign that the calculations are correct. This subject is tangential to our topic (that is, anthropocentric *classifications*, not calculations), but I cannot resist quoting a candid statement by one of the greatest calculators of the century:

> I was thus led to a long calculation, the longest in my career. Full of local, tactical tricks, the calculation proceeded by twists and turns. There were many obstructions. But always, after a few days, a new trick was somehow found that pointed to a new path. The trouble was that I soon felt I was in a maze and was not sure whether in fact, after so many turns, I was anywhere nearer the goal than when I began. This kind of strategic overview was very depressing, and several times I almost gave up. But each time something drew me back, usually a new tactical trick that brightened the scene, even though only locally.
>
> Finally, after about six months of work off and on, all the pieces suddenly fitted together, producing miraculous cancellations, and I was staring at the amazingly simple, final result. . . . Since there were some limiting procedures in my calculation that were not rigorous, I did not feel quite secure until I compared the expansion of the equation in powers of the parameter x, up to x^{12}, with the expansions of Van der Waerden and of Ashkin and Lamb, which were known to be exact to x^{12}. There was complete agreement. (Yang 1983, 12; cf. Yang 1952)

It is clear that Yang's confidence in his calculation is bolstered by the "miraculous cancellations" he experiences after six months. But such thinking is obviously anthropocentric and even childlike: what has "miraculous" simplification of the *notation* we use in describing nature got to do with nature itself?

* * *

"Formalism" is even more blatantly anthropocentric than general Pythagoreanism, because of the pivotal role formalism ascribes to human language.

The ancient attitude is that language not only describes, but mirrors reality, indeed is its own reality.[46] The kabbalists regard the let-

---

[46] As Sidney Morgenbesser pointed out to me, the medieval attitude to language was complicated by their view that natural objects and events were "signs" in a Divine language.

ters of the Hebrew alphabet as the building blocks of creation. One of the characteristics of *magic* is that it systematically substitutes symbols for things symbolized. By erasing a symbol of one's enemy, it is thought, one may eliminate the enemy himself. And the medievals regarded the semantics of language (Latin, Hebrew, whatever) as reflecting the system of "natural kinds." (Plato's *Cratylus* argues this for the Greek language, though some think Plato was joking.)

It was John Locke who "made the modern mind" by arguing that language is conventional, that its semantic schemes need not correspond to a scheme outside language, and that therefore science (itself a linguistic phenomenon) may never fathom the secrets of matter. Unlike Peirce, Locke consoled his readers with the thought that our faculties are sufficient for daily needs.

The idea that we can make scientific progress by studying the *syntax* or other formal properties of our language is even more absurd, on Locke's position. For example, the symmetries of our notation need not reflect the symmetries the notation describes; does a sentence have an enhanced claim to be true because it is a palindrome?[47] To think it does is arrant anthropocentrism. To be sure, the naturalist may appeal to natural selection,[48] which, let us assume, provides an explanation of how love of symmetry promotes survival.[49] But when the preference for symmetries (in food, mating partners, etc.) spills over into an inclination to cherish hypotheses because their *written representation* is symmetrical—we get superstition and magical behavior. (As David Hume used to argue, superstition is as much a subject for scientific inquiry as is rationality. The tendency toward superstition may even have survival value—but it remains superstition, nevertheless.)

[47] A *palindrome* is a sentence that is spelled the same way forward and backward: e.g., "Madam, I'm Adam."

[48] I take it for granted that natural selection is *not* anthropocentric, and that the naturalist may use natural selection in explaining away correlations between humans and their environment, correlations which otherwise would in fact promote anthropocentric feelings among us.

[49] This admission is made for the sake of argument. The naturalist should be careful about such claims, however. Cavemen fashioned symmetric arrow heads for aesthetic reasons, although asymmetric heads would have been more efficient in hunting animals.

Expecting the forms of our notation to mirror those of (even) the atomic world is like expecting the rules of chess to reflect those of the solar system. I shall argue, though, that some of the greatest discoveries of our century were made by studying the symmetries of notation. Expecting this to be any use is like expecting magic to work.

* * *

So far the point is historical, and ad hominem. I assert that physicists behaved as though naturalism were false. But I also assert that they were successful in doing so. And this historical record bears on the cogency of naturalism itself.

If we examine the analogies actually used to discover the major laws of physics in our century, we find that the analogies used are anthropocentric. On naturalist grounds, then, they should have failed, just a dowser should fail to find oil. And this is a difficulty for naturalism, because what the evidence suggests is, on the contrary, that nature looks "user friendly" to human inquiry.[50]

I say "difficulty" (for naturalism) and not "refutation" because it is impossible to refute a background belief like naturalism in this way. We refute a general hypothesis by finding a counter-example (not all crows are black, not all gold is yellow). A background belief operates by labeling certain hypotheses or behaviors as inappropriate or irrelevant. But we know that on occasion, inappropriate hypotheses can be consistent with known evidence, as in the case (discussed above) of the United States Presidents after 1840. The naturalist simply dismisses as a fluke the so-called "success" of these hypotheses, dismisses the (naturalistically) tendentious claim that anthropocentric behavior "worked." (In the same way, where a religious man might say that his prayer was "answered," the naturalist rejects the term.) And, it goes without saying, the naturalist rules out in advance any connection between the human brain and the universe as a whole, except those connections explained away by natural selection.

But if inappropriate hypotheses or behavior continue to "work," at *some* point the naturalist will feel confronted by a mystery: the *apparent* confirmation of that which, according to background

---

[50] I am indebted to George Schlesinger for this phrase.

beliefs, cannot *be* confirmed. As an example, consider the clash between Western and Chinese medicine. For the sake of argument, suppose that Chinese medicine is based on principles unacceptable to Western physicians, so that specific practices based upon it, like acupuncture, are (on Western grounds) doomed to failure. Suppose also, however, that acupuncture, even though on Western principles unconfirmable, has so much *apparent* confirmation that it cannot simply be dismissed. What does one do, short of revising one's total belief in Western medicine to the exclusion of the Chinese?

One strategy is to attempt to explain away the apparent confirmation as an illusion; another is to try somehow to bring acupuncture into the framework of Western concepts. Simply to practice acupuncture while retaining one's background beliefs as before is a kind of intellectual schizophrenia. This is how I regard the position of today's naturalist physicists: their behavior goes counter to their beliefs. Or: given their beliefs, their behavior is irrational. For what seem to be anthropocentric methods of discovering physical laws are so entrenched and widespread and so spectacularly successful that they cannot simply be dismissed.

But because my evidence, like Wigner's, is only evidence of success, there is an obvious counter: ignoring evidence of failure, one can make any hypothesis look good. (This is an example of explaining away the *apparent* success of a hypothesis or procedure.) However, this criticism overlooks the contrast between Wigner's thesis and mine.[51]

My thesis concerns not this or that attempt to *describe* nature mathematically, but the successful deployment of a taxonomy, a scheme. Success is measured by whether the discoveries that scientists were looking for (the laws of atomic and subatomic particles; explanations for the various anomalies of the atomic and subatomic world) were in fact found in due time—using the scheme. It is to be *expected* that many scientists were unsuccessful, though they used the scheme; no classification is a *sufficient* condition for discovery. (If it were, there would be no difference between geniuses and second-

[51] I imply that Wigner's thesis is, in fact, vulnerable to the criticism. In fact, it is not; but my version of the thesis (concerning mathematical analogies) is easier to defend.

rate physicists.) Even for geniuses like Dirac, Yang's rule applies: "In theoretical physics, we are pursuing . . . a guessing game, and guesses are mostly wrong" (Zhang 1993, 19). A system of classification is, then, merely a framework for guessing.

There is, further, a great difference between guessing the right answer on a multiple choice test, where the options are spelled out in advance; and guessing the laws (say) of the interactions between photons and electrons, where the investigator has to make up the options. A scheme of analogies restricts attention to a certain range of options. If the scheme is anti-naturalistic, it should (according to the naturalist) be worthless; and guessing, vain. Peirce's words are so apt that I quote them again:

> 7.680. It is idle to say that the doctrine of chances would account for man's ultimately guessing right. For if there were only a limited number *n* of hypotheses that man could form, so that 1/*n* would be the chance of the first hypothesis being right, still it would be a remarkable fact that man only could form *n* hypotheses, including in the number the hypothesis that future experimentation would confirm. Why should man's *n* hypotheses include the right one? The doctrine of chances could never account for that until it was in possession of statistics of the hypotheses that are inconceivable by man. But even that is not the real state of things. It is hard to say how many hypotheses a physicist could conceive to account for a phenomenon in his laboratory. He might suppose that the conjunctions of the planets had something to do with it, or some relation between the phases of variability of the stars in α Centauri or the fact of the Dowager empress having blown her nose 1 day 2 hours 34 minutes and 56 seconds after an inhabitant of Mars had died.

Even geniuses like Dirac should not have succeeded in guessing the laws of the atom, using the analogical schemes they used, if naturalism is true. This message will be intensified if the reader actually examines Dirac's discovery, narrated below.

The forgetful reader may object that given *later* physics, we often understand why "nonphysical" guesses worked. This objection I have already dismissed: although quarks ground the analogy which led to the "Eightfold Way," my point is epistemological: scientists could not have accepted quarks, for good reasons, when the analogy was made. On the contrary, it was the success of the SU(2)–SU(3) analogy which led scientists to consider quarks: a successful analogy should have a "material" basis. In other words, the successful search for a material basis for the SU(2)-SU(3) analogy, far from discrediting my thesis, provides additional support for it.[52]

The physicist will now expostulate: why *shouldn't* we use mathematical analogies to make conjectures, if we have been so successful with them in the past?

This just shows what is so hard about philosophy, as Wittgenstein used to say: the same misconceptions keep returning again and again—refuted here, they pop up there. What is our physicist saying, after all? That a scientist, in employing a Pythagorean mathematical analogy, is relying not only upon *it*, but also upon the success of *other* Pythagorean mathematical analogies. Then there must be an *analogy* among all "Pythagorean mathematical analogies." We now must ask: does the higher-order property, *being a Pythagorean mathematical analogy,* have a physical basis, at least by present lights? Of course not—this is the same point again, that mathematics as such is an anthropocentric category. Thus, being guided by the success of other Pythagorean mathematical analogies is itself a Pythagorean mathematical analogy, just as anthropocentric as any other.

In sum, on the basis of the evidence about to be presented, I would argue for a weak and a strong conclusion. The weak conclusion is that scientists have recently abandoned naturalist thinking in their desperate attempt to discover what looked like the undiscoverable. This is a conclusion about scientists, not about nature. The strong conclusion is about naturalism: the apparent success of Pythagorean and formalist methods is sufficiently impressive to create a significant challenge to naturalism itself. Now for the evidence.

---

[52] The story of the SU(2)–SU(3) analogy is more complicated than that, however. There are two SU(3) symmetries governing the strong interactions, not one; and Gell-Mann found the "wrong" one. Hence, even quark theory does not explain his success in predicting the omega minus particle.

4

Pythagorean Analogies in Physics

I show that the cardinal discoveries of *contemporary* physics exploited Pythagorean analogies, by analyzing the actual strategies employed by physicists to make those discoveries.

I will begin with strategies which presume that all the solutions of an equation E are akin. This is not by itself a Pythagorean strategy, because often an equation expresses a physical trait which all its solutions exhibit. However, there are cases—the ones I am about to discuss—in which there is evidence that the solutions of a common equation are *not* analogous. A physicist who ignores this evidence, and relies instead on the common equation, pursues a Pythagorean strategy.

> (1) Equation E has been derived[1] under assumptions A. The equation has solutions for which A are no longer valid; but *just because they are solutions of E*, one looks for them in nature.[2] Why is this a Pythagorean *analogy*? A standard way to "derive" a differential equation is to begin with a function f, known already to "be physically real"; and then, by differentiating f, find an equation for which f is a solution. (Usually, there will be several equations like this, so guessing is in order.)

---

[1] By a "derivation," physicists do not usually mean anything like a rigorous deduction from known principles, but plausible reasoning used to "write down" (discover) an equation. I use the term in the same sense.

[2] There are cases, though, where the scientist will rule out solutions as "nonphysical," as Einstein ruled out travel faster than the speed of light, though this is consistent with his equations of special relativity. Such cases are discussed in Steiner 1986.

The assumption that another solution, g, of the equation, is also "physically real" is thus an *analogy* between f and g, mediated by the equation. The analogy becomes Pythagorean if f and g are physically *disanalogous*, so that *only* the equation links them.

The history of the wave idea in physics supplies many good illustrations. Let us consider Maxwell and Schroedinger.

In his *Treatise on Electricity and Magnetism*, Maxwell noted that the experimentally confirmed laws of Faraday, Coulomb, and Ampère, when put in differential form, contradicted the conservation of electrical charge. By tinkering with Ampère's law, adding to it the "displacement current,"[3] Maxwell got the laws to be consistent with, indeed to imply, charge conservation. With no other warrant than this (Ampère's law stood up well experimentally; on the other hand, there was "very little experimental evidence"[4] for the reality of a "displacement current"), Maxwell made the indicated changes. Ampère's law now read that (the "curl" of) the magnetic field is given by the sum of the "real" current and the "displacement current." Ignoring the empirical basis for Ampère's law (magnetism is caused by an electric current), Maxwell now boldly asserted it even for a zero "real" current. This made electromagnetic radiation a mathematical possibility. The belief that it was also physically real required a Pythagorean analogy—one that paid off. Electromagnetic radiation, today the basis of modern communications, was produced in Hertz's laboratory. (You can see why Hertz exclaimed that the mathematical formulas are "wiser than we are.")

An objection can be raised to this "Pythagorean" account, however.[5] Maxwell had already introduced the displacement current in 1862, eleven years before the publication of his *Treatise*, in the course of constructing a mechanical model of electromagnetic phenomena complete with vortices propelling idle (gear) wheels. In order to keep

[3] Ampère's law, as formulated by Maxwell, states that the "curl" of the magnetic field at any point is proportional to the current at that point. Maxwell added the "displacement current," a hypothetical current, equal to the time rate of change of the electric field.

[4] Maxwell 1954, 2: § 608, p. 252.

[5] Enlightening discussions with Jed Buchwald and Alan Chalmers improved my treatment of these issues significantly.

the machine from breaking down under electrostatic conditions, Maxwell had to invest the vortices with electricity, and "the displacement current expressed the flux of the idle wheels owing to progressive elastic deformation of the vortices," as Daniel Siegel points out (Siegel 1991, 86).

The elasticity of the medium implied the possibility of wave disturbances, and Maxwell calculated that the velocity of those disturbances would be the same as the speed of light. This indirectly supported the existence of electromagnetic radiation, because it suggested that light was an example.

This historical background, however, does not change the Pythagorean nature of the 1873 reasoning and the prediction of electromagnetic radiation. Maxwell was a borderline figure, whose thinking became more and more Pythagorean as his theory matured. Maxwell had used his model to derive differential equations of electromagnetism; once he had extracted the equations, the model gradually lost its appeal. The mathematical justification of the displacement current based on charge conservation was isomorphic to the mathematics of introducing it to prevent his model of wheels and vortices from disintegrating (Siegel 1991, ch. 4). The calculation of the velocity of electromagnetic radiation (should it exist) could be done, as Maxwell did it, without assuming an elastic medium. As Siegel points out in great detail (Siegel 1991, ch. 3), in any case Maxwell never believed in the literal existence of the idle wheels. As for the vortices, though they appear in the *Treatise*, where their rotation explains the effect of magnetism on light (Faraday effect), Maxwell later regretted having constructed such a "hybrid" theory of the magnetic effect on light and recommended that younger scientists not follow his lead (Hunt 1991, 18).

The model, being purely hypothetical, gave no support for the existence of electromagnetic radiation as a laboratory phenomenon. This is why Maxwell stated later, in the *Treatise* (quoted above), that there was very little "experimental evidence" for the displacement current (and thus for electromagnetic radiation). The calculation of the speed of electromagnetic radiation as that of light did give circumstantial support that electromagnetic radiation existed (but could also be done independently of the model). Of course, electromagnetic radiation was a possibility, according to Maxwell's equation. But differential equations have many solutions, and there is no reason to believe (particularly in Maxwell's time, when the method of differential equations was not yet standard in science) that we can produce something just because it solves an equationm.

I conclude that Maxwell's reasoning was Pythagorean. By this I mean that once he had a mathematical structure which described many different phenomena of electricity and magnetism, the mathematical structure itself, rather than anything underlying it, defined the analogy between the different phenomena. The analogy, which could be adopted by other physicists (Fitzgerald, Lodge), suggested the existence of electromagnetic radiation (for which there was as yet little evidence) as an experimental phenomenon.

Schroedinger's discovery of wave mechanics also illustrates this strategy.[6] Schroedinger assumed that a particle of constant energy E corresponds to a wave of frequency E,[7] or, more specifically,

$$\Psi(x,y,z,t) = f(x,y,z)e^{-i\frac{E}{\hbar}t}.$$

The equation[8] governing such a wave would be

$$\left[-\frac{\hbar^2}{2m}\nabla^2 + V(x,y,z)\right]\Psi = E\Psi$$

$$\text{where } \nabla^2\Psi = \frac{\partial^2\Psi}{\partial x^2} + \frac{\partial^2\Psi}{\partial y^2} + \frac{\partial^2\Psi}{\partial y^2}$$

formally identical to the equation for a *monochromatic* light wave in a *nonhomogeneous* medium (i.e., where the index of refraction changes from place to place).[9]

---

[6] "Quantisation as a Problem of Proper Values," especially Parts II and IV, in Schroedinger 1978.

[7] This is strictly true only if we set Planck's constant equal to 1.

[8] Note, too, the important fact that even the "space part" of the wave function, namely, $f(x,y,z)$, *also* solves this equation.

[9] The formal (and probably Pythagorean) analogy between the laws of optics and those of mechanics was first pointed out by Hamilton, who called attention to the mathematical similarity between Fermat's principle (law of least time for a light ray) and Maupertuis's principle (law of least action for a mechanical system). Schroedinger considerably broadened the analogy, to include cases where it was impossible to speak of "rays" or "paths." Nevertheless, I hesitate to call the analogy between optics and mechanics Pythagorean *for Schroedinger*, because by 1926 there was evidence supporting a wave theory of matter. Schroedinger expresses great admiration for de Broglie and his wave theories.

To deal with problems in which energy is not conserved—for example, an atom buffeted by a passing light wave—Schroedinger had to get rid of the energy from his equation. Differentiating the wave function once by the time, regarding E as a constant, and using the classical equation[10]

$$E = \frac{p^2}{2m} + V(x,y,z),$$

Schroedinger got

$$\frac{\partial \Psi}{\partial t} = -i\frac{E}{\hbar}\Psi$$

or

$$i\hbar\frac{\partial \Psi}{\partial t} = E\Psi.$$

Substituting in the preceding equation, Schroedinger arrived at

$$i\hbar\frac{\partial \Psi}{\partial t} = \left[-\frac{\hbar^2}{2m}\nabla^2 + V(x,y,z)\right]\Psi$$

(Schroedinger's equation).

Now that energy had been eliminated from the equation, Schroedinger promptly "forgot" his crucial assumption that energy was fixed, and allowed solutions where it wasn't.[11]

For example, Schroedinger's equation allows the superposition of *two* waves, each of different frequency or, what is the same, energy. Yet how can *one* particle be assigned two different energies? (Even today there is no consensus on how to answer this question.) The problem of superposition, by the way, arises even for a temporally constant potential energy field.

Where the potential does vary with the time, the link to the "wave" concept collapses completely. If the potential energy field varies sufficiently (and is not a mere perturbation), the solutions of Schroedinger's equation don't even look like *superpositions* of waves.

[10] As I will point out later, this equation has no meaning in quantum mechanics. Hence, Schroedinger's reasoning raises additional problems, which I will ponder in the last chapter of this book.

[11] Thus, Schroedinger's reasoning has formalist aspects, not merely Pythagorean aspects. Regarding E as a constant in order to derive an equation, and then forgetting that assumption, is a perfect example of allowing the notation to lead us by the nose.

The only remaining "wavelike" characteristic of those solutions is that you don't have a localized particle, but a "smeared out" mass.

Last, but not least, Schroedinger's method (differentiating once by time, twice by space coordinates) creates an equation with an imaginary coefficient, with solutions that have no physical interpretation. A classical electromagnetic wave is simply a state of the electromagnetic field. The wave function in Schroedinger's equation has *complex* values, and in many cases it is not possible to ignore the imaginary part of these values. In the case of the wave of fixed frequency $\Psi(x,y,z,t) = f(x,y,z)e^{-i\frac{E}{\hbar}t}$, by contrast, it is easy to think of $f(x,y,z)e^{-i\frac{E}{\hbar}t}$ as $f(x,y,z)\cos(\frac{E}{\hbar}t)$ (taking the real part). In the general case, the wave function is irredeemably complex; it has no physical interpretation. Only the product $\Psi\overline{\Psi} = |\Psi|^2$ has physical meaning.

In sum, Schroedinger began with a sine wave of fixed frequency, based on an analogy to an optical wave, where the frequency is given by the fixed energy. In writing down the "wave" equation by taking derivatives, Schroedinger completely abstracted away from this intuition, ending with an equation having no parallel in classical optics; one whose solutions had no direct physical meaning; one with superposed solutions; one with solutions having no "wavelike" qualities at all.

Nevertheless, Schroedinger conjectured the equation with all its solutions.[12] Apparently, just being a solution of the Schroedinger equation was enough for belief in its existence. The equation serves as an "umbrella" for all its solutions, and defines all the solutions as "similar." But this is to make the mathematical description as the standard for similarity; which is to say, the similarity is Pythagorean.

Nor can we appeal to our previous experience in physics, where we "wrote down" an equation by beginning with one of its solutions S, and then succeeded in having the equation describe many situations that have little to do with S. For the idea that we can project from our previous success in writing down mathematical descriptions to future successes—this idea is just as Pythagorean as any other, since, again, the only thing common to all those successes is "mathematical"

---

[12] I refer to all "physically admissible" solutions, which for Schroedinger meant "square integrable" solutions.

descriptions. We have no naturalistic definition for "mathematics," or so I have argued.

A final comment on Schroedinger's equation: there are many other equations Schroedinger could have written. In fact, Schroedinger did write down a fourth-order, nonlinear equation (Schroedinger 1978, Part IV, equation (4)), only to reject this real equation in favor of the "imaginary" Schroedinger equation. Why? "In the following discussions," Schroedinger explains, "I have taken a somewhat different route, which is much easier for calculations, and which I consider is justified in principle" (Schroedinger 1978, 104). Though the idea that what is "easier for calculations" is to be preferred is not Pythagorean, it is certainly anthropocentric, as I argued in Chapter 3.

(2) One looks for solutions in nature even where there is reason to doubt their very possibility. There is no *a priori* reason to believe that every solution of an equation has a physical interpretation. There is nothing logically wrong, therefore, with discarding certain solutions of an equation, and it is often done (for example, unbounded solutions of Schroedinger's equation). Nevertheless, the Pythagorean scientist goes by the working hypothesis that a mathematical possibility will be realized by nature.[13]

Take the development of relativistic quantum mechanics. Dirac's equation (whose discovery is detailed in Chapter 6) gave solutions describing the electron better than any previous equation, but it also allowed particles of negative energies as solutions. Dirac, on the contrary, accepted the negative energies as real. But then he had to explain why we never see them. Dirac theorized that negative energy electrons were so ubiquitous as to be undetectable. (There are so many negative energy electrons, in fact, that the "Pauli exclusion principle" prevents the other electrons from joining them.)[14] The electrons we see are the exceptions—they have been "boosted" from their negative energy states, leaving behind a "hole," i.e., an unoccu-

---

[13] This Pythagorean assumption is, it seems to me, a special case of the "Principle of Plenitude," Lovejoy 1964.

[14] The Pauli exclusion principle forbids two electrons from being in exactly the same quantum state. Two photons, by contrast, can be, and often are, in the same state.

pied negative energy state. Such a "hole" would act like a positively charged "electron" with positive energy. And in fact, positrons were discovered.

Today, scientists have a completely different conception of positrons and anti-matter in general. This is, therefore, a different conception of the negative energy solutions. None of this takes away from the spectacular success of Dirac's Pythagorean prediction of the positron.

Another example of this strategy of believing in "impossible" solutions is furnished by the Schwarzschild solution for the equations of General Relativity. This solution describes the gravitational field of a spherically symmetric body, and for bodies of "normal" size and mass, yields not only Newton's laws as a first approximation, but deviations from those laws (such as the precession of Mercury) as a second approximation. But if the body is smaller than its "gravitational radius" (for a body the mass of the Sun, the gravitational radius is about 3 kilometers), one of the components of the metric tensor goes to infinity at a distance of the gravitational radius from the center of the body. This seemed to rule out such a dense body, as it would create a "singularity" in spacetime itself. Nevertheless, scientists had enough faith in the equation to believe in even this solution; in 1954, Finkelstein argued that the singularity at the gravitational radius was not a real singularity of spacetime, but an artifact of the coordinate system.[15] From a distant vantage point, a spherical mass smaller than its gravitational radius would appear as a "black hole" (an idea which, in the context of classical gravitation, goes back to Laplace at least);[16] a test particle falling toward the mass would seem to "us" to take infinitely long to get to the gravitational radius, but an observer on the test particle would not notice anything special as he crossed the gravitational radius (the "Schwarzschild sphere"). Scientists are now persuaded of the actual existence of black holes.

[15] For an elementary example, consider the "origin" in polar coordinates. Though in Cartesian coordinates, the "origin" has a straightforward representation as (0,0), in polar coordinates, this same point can be represented by infinitely many points of the form $(0,\theta)$, for any angle—which obviously affects the mathematical properties of functions defined on the plane. Obviously, this singularity has no physical meaning whatsoever.

[16] Thanks to Barry Simon for pointing this out.

A final point: the search for solutions in nature may require reinterpreting the solution or the equation. Quantum mechanics is an extreme example: the equations were "believed" even before anybody "knew what they meant." This faith in the formalism as being "wiser than we are"[17] is what motivates higher-order mathematical analogies.

(3) Suppose we have successfully classified a family of "objects" by a mathematical structure S. Then we project that this structure, *or some related mathematical structure* T, should classify other families of objects, even if, given present knowledge, (a) S is not reducible to a physical property, and (b) the relation between S and T is not reducible to a physical relation. We have doubly Pythagorean analogies.

This reasoning has been rampant in elementary particle physics, where "symmetry arguments" have led to some remarkable discoveries. These are Pythagorean arguments from analogy, where one symmetry is derived from another. According to the modern definition, an object has a symmetry if it is invariant under a group of "transformations." For example, a ball has rotational symmetry, since any rotation leaves it as it was. The sun's gravitational field has rotational symmetry, though the (elliptical) orbits of the planets do not.

The importance of symmetries in physics lies in their relationship to laws of conservation. Each symmetry of a physical system implies a law of conservation[18]—rotational symmetry, for example, implies the conservation of angular momentum. In fact, for the conservation of angular momentum, all that is necessary is that the system in question be invariant under "infinitesimal" rotations.[19]

During the twenties, the electron was discovered to have an

[17] Hertz, quoted above.

[18] For continuous symmetries, this is a theorem proved by the mathematician Emmy Noether, and it is true both for classical and quantum mechanics. In quantum mechanics, the theorem is rather obvious—since the physical magnitudes are actually *identified* with symmetry transformations. (To be more precise, every magnitude is represented by an operator that is proportional to the corresponding symmetry transformation.) For this reason, the connection between symmetries and conserved quantities in quantum mechanics is not limited to continuous symmetries—for example, right–left symmetry is equivalent in quantum mechanics to conservation of parity.

[19] This has to do with the mathematical properties of the algebra of infinitesimal rotations more than with the physical properties of angular momentum.

"intrinsic" angular momentum, called "spin." Namely, although the electron cannot be regarded literally as rotating,[20] it sometimes acts as though it were. For example, a spinning charged ball is a magnet, with the north and south poles determined by the axis and the direction (clockwise or counterclockwise) of its rotation. The electron turned out also to be a little magnet.

But if we test the electron by placing it between the poles of a horseshoe magnet, we find a mystifying difference between it and the classical charged rotating ball: the electron is always found aligned so that its north pole is pointing directly at the north pole, or the south pole, of the horseshoe magnet! This of course is not true of a classical spinning charged ball, which can have its north pole pointing any which way *initially* upon measurement. The usual way of putting this is that every electron, upon measurement, can be in one of two possible spin-states, which are thus called "spin up" and "spin down." (Prior to measurement, the electron may be in a "superposition" of the two states.)

Here is another, even more, mystifying feature of electron spin.[21] Suppose we have an electron placed between the poles of a very strong horseshoe magnet, with the north pole of the electron pointing to the north pole of the magnet. If we rotate the horseshoe magnet very slowly around any axis, the north–south pole of the electron will follow the magnet, so that at any moment, if we make an observation, we find that the north pole of the electron is still pointing at the north pole of the magnet.[22] If we turn the magnet 360 degrees, you might think that the electron should return to its initial position—and it does, almost. The electron is once again oriented with its north pole pointing "upward," yet the rotation it has suffered has "marked" it in a very subtle way: its state vector is multiplied by -1.[23]

---

[20] See Appendix A.

[21] Thanks to Shmuel Elitzur for helping me with this material.

[22] If we rotate the magnet fast, or if the magnet is weak, there is a possibility of the electron flipping upside down, with the north pole pointing at the south pole of the magnet.

[23] Thus, to really get the electron back to its *status quo ante*, you have to rotate it 720 degrees. Though there is no way to imagine what is going on here, Feynman used to act out this kind of symmetry with a coffee cup—you can turn a full coffee cup 360 degrees while holding it in your hand, without spilling a drop, only your arm gets twisted. Surprisingly, you can turn the cup another 360 degrees (try it), and your arm untwists itself. (Dirac did this stunt with belts.)

Now this change, being a *uniform* change of coordinates, has no physical significance, as we have already seen, so long as we are talking about a single electron. But suppose we have two electrons: one so rotated and one left alone. Then, astonishingly, the two electrons *interfere* with one another: for example, if we send a beam of electrons through two slits so that they hit a screen, and install a "rotating" mechanism in front of one slit so that the electrons going through it undergo a 360 degree rotation as above,[24] then there will be a dark spot on the screen midway between the slits.

There are thus two characteristic features of electronic spin: the duality of the spin states (up and down), and the changing of the sign at 360 degrees. It is easy to represent this situation mathematically (so long as we don't try to imagine what is going on): the spin state of the electron is represented by a two-dimensional complex vector (one dimension for "up", one for "down"), and the effect of a spatial rotation of the electron is to transform the vector by a certain 2×2 complex matrix, the matrix corresponding to 360 degree rotation in any direction being $\begin{pmatrix} -1 & 0 \\ 0 & -1 \end{pmatrix}$. The group of such matrices is called SU(2).[25] In making this representation, we are of course ignoring everything about the electron but its spin. Finally, because of the connection between symmetry and conservation, spin is conserved in any physical system which remains physically invariant when we apply an SU(2) transformation to it.

In 1932, Heisenberg conjectured boldly that the proton and the neutron—ignoring their opposite charge—are two states of the same particle, "spinning" in opposite directions in a fictitious three-dimensional Euclidean "space." The space had to be fictitious, since (unlike the situation with the "up-down" electronic states) one cannot turn a neutron into a proton by standing on one's head. Heisenberg reasoned that the nucleus of the atom is invariant under SU(2) transformations, those which describe the spin properties of the electron; and that there had to be, therefore, a new conserved quantity, *math-*

[24] The mechanism has to be such as to leave no traces of its activity, otherwise we would know through which slit the electron passes and the interference effects are annulled.

[25] Technically, SU(2) is the group of all 2×2 unitary matrices with determinant $+1$. A unitary matrix is one that does not change the "length" of the unit vector, hence is analogous to rotation in Euclidean space.

ematically analogous to spin. This quantity is today called isospin, and its discovery launched nuclear physics. Both the neutron and proton, states of the same particle, are called "nucleons."

Note that Heisenberg's theory was not *just* that the neutron and the proton are the same particle. That hypothesis would require only a weaker symmetry: that one could "swap" neutrons and protons discontinuously (the permutation group), in any physical process not involving the charge. Heisenberg's theory is that the neutron is obtained from a proton by a continuous *abstract* "rotation of 180 degrees," and also that to return a neutron or a proton to its initial isospin state, one must "rotate" the particle a full 720 degrees in the *fictitious* isospin space. It seems clear that the mathematics is doing all the work in this analogy, and that Heisenberg's analogy was highly Pythagorean. (Indeed, even today, nobody knows why electron spin and isospin have the same symmetry—and even if someone were to explain the coincidence, the explanation was not available to Heisenberg in 1932.)

At this point, it is better to stop altogether relying on visual intuition of three-dimensional space, fictitious or real, and to think of the SU(2) group itself as the fundamental descriptive idea. The SU(2) transformations are, as we said, 2×2 matrices. The dimension of these matrices, two, corresponds to the two "states" of spin/isospin. Spin and isospin occur in pairs (proton/neutron, electron-up/electron-down). But group theory shows that matrices of any dimension n can "represent" the SU(2) group, and the physical meaning of this is that isospin (as ordinary spin) can come in groups of any whole number n: triplets, quadruplets, quintuplets, etc.[26]

The power of SU(2) symmetry was demonstrated in 1938, when Nicholas Kemmer reasoned, as in the preceding paragraph, that there could be an isospin triplet (a particle capable of being in one of three isospin states)—just as in the case of electron spin—and predicted the properties of the three *pions* nine years before the experimentalists were able to verify them. Like the nucleons (neutron and

[26] The magnitude of the isospin of the nucleon is ½ Planck's constant. In general, the relation between the magnitude of the isospin, and the number of its states is given by $n = 2s + 1$, where s is in units of Planck's constant. For the nucleon, therefore, $s = \frac{1}{2}$; $n = 2$ (two states, the proton and the neutron). For the pion, $s = 1$; $n = 3$. For the delta, $s = \frac{3}{2}$; $n = 4$, etc.

proton), the pions exert the "strong" nuclear force. Similarly, the deltas form an isospin quadruplet. Kemmer's analogy was simply an extension of Heisenberg's, and equally Pythagorean.

Here is a truly stunning example which shows the power of Pythagorean analogies using SU(2) symmetry.[27] It is an example of two different—but mathematically equivalent (isomorphic)—systems, both of which are known in advance—unlike the previous two examples—to have SU(2) symmetry. The physicist projects that there ought to be a physical process which can transform one system into the other. The heuristic rule here is that a mathematical isomorphism betokens physical equivalence. Because this projection is not deductively required even by the precepts of quantum mechanics, the projection is clearly Pythagorean. Here are the details.

Given two linear spaces M and N, of dimensions m and n, mathematicians "add" and "multiply" the spaces to form spaces of dimensions $m + n$ and mn.[28] If a vector of M is of the form (a,b,c) and that of $N,(d,e)$, we form the five-dimensional space M⊕N by creating all vectors of the form (a,b,c,d,e). The "product" of M and N, which has six dimensions, is the vector of form (ad,ae,bd,be,cd,ce).[29] An important difference between M⊕N and M⊗N is that there is a canonical embedding of M and N in M⊕N; not so for M⊗N.

In quantum mechanics, the interpretation of these two operations is as follows: if vectors Ψ_M in M and Ψ_N in N describe states A and B, then $\Psi_M \oplus \Psi_N$ is the superposition of the two states—in which either A or B is possible. Should A and B be distinguishable states, the product $\Psi_M \otimes \Psi_N$ describes the pair state <A,B>, where both A and B obtain.

Let us apply this to particle physics. We recall that physicists (after Heisenberg) regard the proton and neutron as two states of the same particle, the "nucleon." The same is true for the three pions (positive, neutral, and negative) and the four delta particles (double-positive, positive, neutral, negative). If we fix all other physical magnitudes (such as momentum, spin, etc.), we can think of the nucleons as defining a two-dimensional vector space, the pions as defining a

[27] See Sternberg 1994, 4.3, 4.8 (from which the present account is adapted), for both mathematical and physical details of this example.

[28] I refer to the *direct* sum and *tensor* product of linear spaces.

[29] For rigorous details, see Sternberg 1994, Appendix A.

three-dimensional vector space, and the deltas as defining a four-dimensional space. (But recall that these vector spaces are complex spaces, meaning that the coordinates of the vectors are complex numbers.)

Each of these particles, nucleons, pions, and deltas, have SU(2)-symmetry. This means that any fact (equation) about these particles remains true when we apply any SU(2) transformation uniformly to each particle.

Up to this point, there is nothing new.

But now consider the following mathematical fact: there is an isomorphism between

> the delta space (four dimensions) plus the nucleon space (two dimensions) on the one hand; and

> the pion space (three dimensions) times the nucleon space on the other.

Let us call the former, the sum space; the latter, the product space.

Furthermore, this isomorphism is an SU(2)-preserving isomorphism. That is, if a state x in the sum space corresponds under the isomorphism to a state y in the product space, and g is an SU(2) transformation, then $g(x)$ corresponds to $g(y)$ under the isomorphism.

Again, we have an interesting fact of pure mathematics. But now comes the Pythagoreanism. Given the (SU(2)-preserving) isomorphism between the product space and the sum space, the physicist assumes that there is also a physical equivalence. Thus, there must be an experiment which transforms a pion-nucleon pair into a *superposition* of a delta and a nucleon.

In a nutshell: because the sum is mathematically equivalent to the product, we regard them as *physically equivalent*.

This reasoning does not even follow from the rules—strange enough in themselves—of quantum mechanics. The product of the nucleon space and the pion space describes a system with two *actual* particles: a nucleon and a delta. The sum of the nucleon space and the delta space describes a system with only *one* particle, which is *either* a nucleon or a delta. There is no rule connecting these spaces. The most we can say is that the physical equivalence of the two systems is possible, because by SU(2) symmetry, every equivalence of the hadrons has to be SU(2) preserving, and we have seen that the isomorphism between the sum space and the product space is, in fact, SU(2) preserving. But what is possible need not be actual.

And yet the reasoning works spectacularly: if we make pions collide with nucleons at the right energies and momenta,[30] we can actually create the superposition of a delta and a nucleon. The details of this are subtle, so I will relegate them to Appendix B. The intrepid reader who works through the Appendix will be much more mystified than one who doesn't: theorems of group theory, and nothing more, allow detailed numerical predictions which appear to come out of thin air, though following mathematically from the Pythagorean hypothesis:

$$\text{mathematical equivalence} = \text{physical equivalence}.$$

* * *

Isospin symmetry was nonspatial, but isomorphic to spin symmetry. The analogy is certainly Pythagorean—indeed, even today, physicists see no *physical* analogy between the quantities "spin" and "isospin," and therefore have no explanation for the success of Heisenberg's reasoning.

Physicists continued to introduce, by Pythagorean analogy, formal symmetries whose relation to experience was increasingly tenuous. An example is the discovery of "unitary spin,"[31] also called the "Eightfold Way" because unitary spin has eight "components."[32] This scheme, discovered independently by Gell-Mann and Ne'eman, aimed to incorporate a newly discovered conserved quantity— "hypercharge," not discussed here—and isospin into a "global symmetry," known as SU(3).

[30] The Pythagorean reasoning described here does not predict the actual energy at which this will happen. To do that, we would have to know more details about the nature of the interaction among the particles—whereas in fact all we are given are the symmetries of the interaction.

[31] This term, introduced by Gell-Mann, is not used today.

[32] Isospin, like spin, has three components, since we can have spin in the direction of the x-, the y-, and the z-axis—only one of which, by the Uncertainty Principle, can be measured at a time. Unitary spin adds five more components to these three, of which at most two can be simultaneously ascertained.

Like isospin, the theory of unitary spin postulated that strongly interacting particles that are prima facie different may actually be different states of the same "thing." Thus, the strongly interacting particles may be divided into families whose members share a common unitary spin[33] but are nevertheless in different unitary spin "states." Again, this classification flows mathematically from a symmetry property of strong interactions—namely, that they are invariant under a transformation group, as in the case of isospin.

But unitary spin presents also interesting novelties. First, the number of particles in a unitary spin family is restricted. Unlike spin or isospin, where families can have any number of members, unitary spin families can have only 3, 8, 10, 27, ... members—the series being determined by algebraic considerations. In fact, Gell-Mann and Ne'eman made a spectacular prediction of the "omega-minus" particle by noting that nine known particles could belong to a "decuplet" of unitary spin, provided that the missing tenth particle existed.

Second, the "Eightfold Way" scheme grouped together particles for which there was little evidence that they were "different states of the same thing." The particles would have to have the same mass/energy, yet—for example—in the "octet" family of the Eightfold Way, the heaviest particles were 50 percent heavier than the lightest (which were the proton and neutron).[34] What actually turned out to be the case, that unitary symmetry is "broken" by a natural effect that disguises the underlying order, might have looked like an ad hoc attempt to save the theory. Thus the discoverers were gripped by a strong faith in the symmetry of the basic forces of the universe.

But most significant is this disparity between unitary spin (or SU(3)) and isospin (or SU(2)): the group SU(2), though not isomorphic to the group O(3) of rotations in three dimensions, is two-to-one homomorphic to it (which is why we said it takes an electron two rotations to get back to its original state). SU(2) is called by mathematicians the "double covering group" of O(3). Furthermore, in the infinitesimal limit, the homomorphism becomes an isomorphism.[35]

---

[33] The unitary spin of a system is defined by an ordered *pair* of numbers, giving the isospin and the hypercharge of the system.

[34] Compare the isospin families: the mass difference between proton and neutron, for example, is one part in seven hundred.

[35] In mathematical jargon: the *Lie algebra* SU(2) is isomorphic to the *algebra* O(3).

(This is why we can treat electronic spin as a form of angular momentum.) Historically, the matrices of SU(2) were used to represent physical rotations.

Unitary spin symmetry, or SU(3), by contrast, is not isomorphic, even in the infinitesimal limit, to rotations in any dimension.[36] Unitary symmetry is just an abstract symmetry—invariance under a transformation of a three-dimensional *complex* space.[37] When we get to SU(3), the link with perception has been snapped. And this makes the SU(3) hypothesis, the analogy to SU(2), grossly Pythagorean.

How did Gell-Mann discover the Eightfold Way?[38] In particular, was the analogy to isospin physical or Pythagorean? Had he known then about quarks, which are the "triplets" of unitary spin, he could have built up the strongly interacting particles from quarks, just as Heisenberg built up the nucleus from nucleons (which are the fundamental "doublets" of isospin), without any arguments from analogy.

But in 1960 there were good reasons to deny quarks, because particles with fractional charge had never been observed. Even after Gell-Mann himself proposed quark theory, suggesting that they might be unobservable, he was attacked by Marxist physicists for "bourgeois idealism."[39] What the Marxists had in mind, presumably, were passages like these:

> [W]e construct a mathematical theory of the strongly interacting particles, which may or may not have anything to do with reality, find suitable algebraic relations that hold in the model, postulate their validity, and throw away the model. We may compare this process to a method sometimes employed in French cuisine: a piece of pheasant meat is cooked between two slices of veal, which are then discarded. (Gell-Mann and Ne'eman 1964, 198)

[36] That is, the algebra SU(3) is not isomorphic to any algebra O(n).

[37] For details, see Gell-Mann and Ne'eman 1964, which contains the original papers.

[38] What follows is based on Gell-Mann's recollections in Gell-Mann 1987, and correspondence with Yuval Ne'eman. I reiterate: I discuss here only Gell-Mann's reasoning, because Ne'eman's route to the Eightfold Way, described in the same volume (Doncel 1987, 499–510), did not involve mathematical analogies.

[39] See Gell-Mann 1987, 494. These are the physicists under the leadership of Sakata who had championed a "triplet" model of their own, in which "materialistic" protons, neutrons, and lambda particles make up strongly interacting particles.

In fact, Gell-Mann was led to the Eightfold Way by a tortuous road. He drew his third-order mathematical analogy without knowing it and in the "wrong" way.

Gell-Mann was attempting to generalize the Yang–Mills equations,[40] which were the first, though failed, attempt to write down an analogue of Maxwell's equations for the nuclear field. The field itself has "local isospin symmetry": its behavior is invariant under independent rotations of the isospin at every point in spacetime. (Thus, for example, what is called a proton at one point may be called a neutron at another. This is a much stronger symmetry, therefore, than Heisenberg's discovery.) Gell-Mann saw that Yang–Mills theory provides a recipe for writing a field equation, given merely the appropriate local symmetry;[41] hence, to generalize the equation, one must generalize the concept of isospin. What Gell-Mann did without knowing it was to characterize isospin rotations as a "Lie Algebra"— a concept reinvented for the occasion, but known to mathematicians since the nineteenth century. He then (by trial and error) began looking for Lie Algebras extending isospin—unaware that the problem had already been solved by the mathematicians—but failed, not realizing that the first solution required eight components, as above.[42] Later, a mathematician at Caltech enlightened him on Lie Algebras. Gell-Mann's was indeed a Pythagorean analogy, if dimly understood.

Gell-Mann was also "lucky." The Yang–Mills equations were designed to describe the nuclear (or strong) interaction. And what Gell-Mann called unitary spin is not a property of that interaction. In other words, the property of quarks responsible for their mode of interaction (called "color" today) is different from the property ("flavor") that determines their "unitary spin" state. Luckily for Gell-Mann (and for science), "color" and "flavor" have the same symmetry,[43] a coincidence for which no explanation is known.[44] So

[40] Yang and Mills 1954; reprinted in Yang 1983. The derivation of these equations is discussed in detail below, Chapter 6.

[41] "From S-matrix to Quarks," in Doncel 1987, 489.

[42] He stopped at seven, exhausted by bouts of wine drinking (Doncel 1987, 489).

[43] Except, of course, that the "color" symmetry is a local symmetry, valid at every point of spacetime; whereas "flavor" is a global symmetry.

[44] Compare, for example, the electron, which is observed always to be in one of two possible states ("up" and "down"). The (global) symmetry (SU(2) or spin symmetry)

even the claim that Gell-Mann's success in arguing by mathematical analogy is "explained" by the theory of quarks must be attenuated, since Gell-Mann discovered the "wrong" symmetry.

It might be argued that, even without an underlying physical base, the use of group theory in physical discovery is empirically justifiable, because group theory is grounded upon symmetry, an empirical notion. But this is circular reasoning, since "to say that an object has a symmetry just means that it admits a transformation into itself, and the collection of all these symmetries is a group."[45] The groups invoked in the theory of elementary particles today express symmetries only in this question-begging sense; they do not express empirical or geometrical symmetries. The analogies here were therefore, in fact, Pythagorean. In fact, even the *discovery* of quarks might be regarded as Pythagorean, although the existence of quarks *explains* the isopsin symmetry. The success of the abstract classification of the particles by SU(3) made it "natural" to look for three basic particles, by analogy to SU(2), where we have the proton and neutron as the basic particles. "Natural"—for the Pythagorean.

* * *

We have discussed solutions and symmetries of the laws of nature—now, we examine the equations themselves.

(4) One formulates equations by analogy to the mathematical form of other equations, even if little or no physical motivation exists for the analogy.

A case is Einstein's derivation of the field equations of General Relativity. His method was to set down three mathematical conditions that the equations should satisfy—then he proved that essentially only one equation satisfied them. We are interested here in two of Einstein's conditions: that the equation should be a second degree

---

responsible for this classification is not the same as that of *local* symmetry (U(1)) which governs the interaction of two or more electrons. That strongly interacting particles are classified by the same (global) symmetry as the (local) symmetry of their interactions with one another, is one of the great lucky breaks of the history of science—and a great triumph of Pythagorean "abduction."

[45] Chandler and Magnus 1982, 52–3.

differential equation, and linear in the second derivatives. Where did he get these conditions?

Einstein himself says that the conditions that no more than the second derivatives of the metric tensor should appear in the field equation, and that the equation be linear in the second derivatives, both were "naturally taken from Poisson's equation."[46] Poisson's equation is the nineteenth-century form of Newton's law of gravitation, which indeed involves only the second derivatives of the "gravitational potential," and those only linearly. It is reasonable that a scientist would try to have this equation as a special or limiting case of any future equation of gravity, such as General Relativity.

This condition, though, does not imply that GR must have all the mathematical properties of Poisson's equation. GR could be a higher-order equation, or not linear in the second derivative, and still imply Poisson's equation as a limiting case. Therefore, when Einstein says that the mathematical properties of GR were taken from Poisson's equation, he means that the coordinates of the metric tensor "play the role" *mathematically* of the gravitational potential of the Poisson equation. Einstein's original paper on GR contains no real physical argument for this analogy,[47] and Graves agrees with me that there wasn't any.[48] Thus, the analogy with the Poisson equation was a Pythagorean analogy.

Another example of this type of induction—the derivation of an equation from another one, using a Pythagorean mathematical analogy, is the procedure of Heisenberg (with Born and Jordan) in deriving matrix mechanics.[49]

Heisenberg began with the classical Hamiltonian equations of mechanics, and substituted matrices for the variables appearing in that equation. In Hamiltonian mechanics, for each system we construct a function $H(q,p)$ of the coordinates q and the momenta p—this is typically a polynomial in the qs and ps. The function H is called the *Hamiltonian* of the system, and reflects its specifics: spherical symmetry, an inverse square law, etc. Now we can construct differential equations of the form

---

[46] Einstein 1974, 84. [47] Ibid., 79–81.

[48] Graves 1971, 178. Graves asserts explicitly that the analogy was formal.

[49] Heisenberg 1925; Born, Heisenberg, and Jordan 1925.

$$f(H) = \frac{\partial}{\partial t} F(q,p,t)$$

which give the time dependence of a function F of the coordinates, momenta, and the time, as the system sweeps deterministically through time and space, as a functional f of the Hamiltonian.[50] The functional f will depend on the function F we want to calculate, but in all cases will contain differential operators.[51] (To simplify matters, I shall further abbreviate any Hamiltonian equation of this type as $E(H)$.)

Now Heisenberg's idea—as amplified by Born and Jordan—was as follows. Given any quantum system which would have, if a classical system, Hamiltonian H, and be described, therefore, by an equation $E(H)$, simply replace all the variables p, q in the equation by *matrices*, and all operations (addition, multiplication, differentiation) by corresponding matrix operations. The result is a matrix equation $E^*(H^*)$, and this is the quantum equation which governs the system.[52]

This procedure is an example of "quantization"[53]—i.e., of transforming a false classical equation for an atomic system into (what is hoped to be) a true quantum equation for that system. There is no physical rationale for this procedure, that of substituting matrices for variables, except a Pythagorean analogy. To put the matter another way, it is impossible to imagine a physicist *discovering* the matrix equation by direct physical reasoning, skipping entirely the classical step. This is because the matrix equation, though one can extract measurable "numbers" from it, and therefore confirm the equation, does not "say" anything about the physical system which can be

[50] We think of the system as a single point in a multidimensional phase space, with one dimension for each coordinate and each momentum.

[51] Thus, if all we want is the time dependence of the second space coordinate, so that F is nothing but a trivial "projection" function, then all we need to do is take the partial derivative of H with respect to the second momentum coordinate. In general, the story is much more complicated, and involves the "Poisson bracket" of F with H. Cf. Goldstein 1950, chs. 2, 7.

[52] The knowledgeable reader will note here that this description of quantization is ambiguous, because in matrix multiplication, AB may not be the same as BA, whereas functions in classical mechanics are all commutative. A discussion of this problem, which only increases the "mystery" of quantization, will be found in Chapter 6.

[53] See Chapter 6 for more on quantization.

expressed—even qualitatively—without the matrices. Matrices as such have no independent physical meaning. The matrix equation is parasitic on the false classical equation, which does have a "non-mathematical" meaning: the Hamiltonian equation expresses the conservation of energy of the classical system. The matrix equation does not express any such thing (as we shall see later, conservation of energy is literally meaningless for a single quantum system; we can speak only of the conservation of the expectation values of the energy), though, again, one can manipulate the matrix solutions to extract information about energy. Matrices don't even behave like ordinary numbers, because although matrix "multiplication" is associative, it is not commutative: $AB \neq BA$, in general.

For this reason, the success of the Heisenberg (Born and Jordan) strategy in one case would not increase the likelihood of its success in another—*unless* we take Pythagorean analogies seriously. Conversely, if we reject Pythagoreanism, then we are obliged to treat each new substitution of matrices for variables as independent of the others. The mathematical analogy between the matrix and the classical equation is just that: mathematical.

For it is necessary to keep reminding ourselves, that the success of a strategy in one instance does not affect the likelihood of success in another, unless it's the "same" strategy in both cases. And what is the "same" is heavily dependent on the way we are disposed to judge similarity, based on our background beliefs. I am arguing that the background beliefs at work here in the Heisenberg (Born and Jordan) strategy are Pythagorean; the mere reliance on this strategy—the mere judgment that substituting matrices in one equation is the "same thing" as substituting them in another—even in the context of guessing or "abduction," betrays a Pythagorean bent. And, if I am right that Pythagorean reasoning is anthropocentric, we must admit that one of the greatest discoveries in physics this century was made by "non-physical" or anthropocentric reasoning.

Can the Correspondence Principle of Bohr help? The usual interpretation of the principle is that the classical equations are to be the limiting case of the quantum equations, as the systems described get larger and larger with respect to Planck's constant $\hbar$—alternatively, as we formally allow $\hbar$ to approach zero. And this principle is satisfied by the Heisenberg strategy. But, clearly, the Correspondence

Principle—so interpreted—does not determine the form of the quantum equations.

There is an interpretation of the Correspondence Principle—Dirac's—which *is* relevant to Heisenberg's procedure.[54] After rediscovering, independently, a number of the insights of Born and Jordan, and adding to them,[55] Dirac concluded: "The correspondence between the quantum and classical theories lies not so much in the limiting agreement when $\hbar \to 0$ as in the fact that the mathematical operations on the two theories obey in many cases the same laws."[56]

It might be objected that Dirac, but not the other Founding Fathers of quantum mechanics, was in fact a Pythagorean. Yet a recent history of quantum mechanics (Darrigol 1992) analyzes the Correspondence Principle, case by case, to show that the applications of the Principle—by Bohr and many others—in every case were formal (what I call here Pythagorean) analogies. I can only refer the reader to Darrigol's lucid exposition.

* * *

Another equation derived by Pythagorean analogy is the Klein–Gordon equation, a relativistic version of Schroedinger's—in fact, Schroedinger published it before Klein and Gordon.[57] Schroedinger noted that his (nonrelativistic) equation could be obtained from the classical energy-momentum relation

(CE) $$E = \frac{p^2}{2m}$$

by formally substituting differential operators for E and p. Now the corresponding relativistic equation is

(RE) $$E^2 = p^2c^2 + m^2c^4.$$

Schroedinger thus suggested making the "identical"[58] substitution

---

[54] Dirac 1926, reprinted in Van Der Waerden 1967.

[55] Dirac discovered the formal analogy between the classical Poisson brackets and the commutator of two operations or matrices in quantum mechanics.

[56] Dirac 1926, 315. See also Chapter 6.

[57] Schroedinger 1978, Pt. IV, pp. 118–20. Pais 1986 lists six authors who derived this equation in the space of half a year, in 1926.

[58] I remind the reader here that the notions of "identical" and "doing the same

for E and p as in the classical case, and obtained the Klein–Gordon equation.[59]

What motivated these substitutions? Manifestly, this argument: since substituting certain operations for magnitudes transforms non-relativistic mechanics to quantum mechanics, the very same substitution must transform relativistic mechanics to relativistic quantum mechanics. Since no physical argument justifies this, we conclude that the argument is by a Pythagorean analogy. Indeed, Schroedinger himself tells us that his relativistic equation is based on a "purely formal analogy."[60]

True, Schroedinger expresses "the greatest possible reserve" about his relativistic equation.[61] And one reason he gives is the lack of

---

thing again" are far from transparent. On the contrary, it is our classificatory scheme that determines what we consider "doing the same thing again." My claim is that Schroedinger, whether consciously or not, was employing Pythagorean analogies which not only supported his behavior, but also determined that what he was doing was the "same" as in the non-relativistic case.

[59] More precisely, Schroedinger did not write down the Klein–Gordon equation in this form, which is an equation for a free particle. For, according to that equation, the probability for the particle to be in a certain region can be negative, i.e., meaningless. Instead, Schroedinger multiplied the probability density function by the electric charge e, interpreting it as a charge density function—charge density, unlike probability density, can be negative. And his equation was derived by substituting in the relativistic energy-momentum equation for an electron in an electromagnetic field, rather than in that of the free particle.

[60] Schroedinger 1978, 118. Schroedinger repeats this phrase on p. 119.

When Linda Wessels (Wessels 1977, 328) states that Schroedinger had arrived at the "Klein–Gordon" equation before he published his four-part series on wave mechanics, she must have had the (abortive) relativistic amplitude equation (a time-independent equation) in mind. For she herself states, on the previous page, that Schroedinger at this time was attempting to treat the hydrogen atom as a "vibratory" problem involving standing waves. The sources she cites (328 n.) refer vaguely to failed relativistic attempts, but do not cite the KGE by name. To settle the matter, I examined (microfilms of) Schroedinger's notebooks. Sure enough, my search turned up the (unpublished) relativistic amplitude equation, but not the "real" Klein–Gordon equation, which is a Lorentz-invariant, time-dependent equation (see Steiner 1989, 470 n.).

Even supposing, however, that Schroedinger had arrived at the "real" KGE in 1925, he also knew that it misdescribed the electron. His faith that it nevertheless describes *something*, and his willingness to publish it in 1926, he based solely on the formal analogy (for he published no other reasons for accepting the equation). Schroedinger's faith was vindicated years later, when it was discovered that the Klein–Gordon equation (in quantum field theory) describes particles not known to exist in 1926—pions.

[61] Schroedinger 1978, 119.

physical basis for his derivation. So one might suspect Schroedinger of weak faith in the applicability of mathematics. But his real objection, as he himself points out, is that the equation does not yield the correct energy levels of the hydrogen atom; these are influenced by the spin of the orbital electron, and the solutions of the Klein–Gordon equation do not have "spin." Charged particles with no spin were not at that time known. On the contrary, then: Schroedinger had such faith in his formal analogy that he was willing to publish it, even though he had not worked out a single application of the Klein–Gordon equation.

Another problem with the Schroedinger interpretation of the Klein–Gordon equation was the existence of "negative energy" solutions, just as in the Dirac equation. Dirac's treatment of negative energy solutions, described above, could not work here, because the Pauli exclusion principle is not true for spinless particles.

It took till 1934 to "understand" the Klein–Gordon equation, when Pauli and Weisskopf interpreted it as a field equation rather than a particle equation. In another application of inductive pattern (4), they applied to the Klein–Gordon equation the so-called "field quantization technique," which had been used to extend Dirac's equation for the electron to a many particle theory, able to describe the creation and destruction of material particles. In the case of the Klein–Gordon equation, however, Pauli and Weisskopf began with an equation that had *no* physical solutions, and derived a field equation with, finally, a positive energy parameter. This equation[62] does, in fact, describe spinless, charged particles: for example, the positive and negative pions, discovered subsequently. Our story, then, doubly illustrates pattern (4): in the transition by analogy from Einstein's equation to the Klein–Gordon equation, and from the Klein–Gordon equation to its field-theoretic extension.

Indeed, the mathematical analogy of the Klein–Gordon equation to Einstein's equation was so compelling that Dirac, despite his rejection of the Klein–Gordon equation,[63] used it in deriving his own

---

[62] I am referring to the *complex* Klein–Gordon equation.

[63] Cf. Pais 1986, 288–92. Dirac objected most of all to the second time derivative in the Klein–Gordon equation; see Dirac 1958, § 17. Dirac presents a "proof" that quantum mechanics requires a first order time derivative.

equation for the electron. Having derived the general form of his equation, but lacking the coefficients, Dirac pinned down the coefficients by requiring that any solution of his equation would also have to be a solution of the Klein–Gordon equation. Whatever was wrong with the Klein–Gordon equation, he argued, it at least expressed the relativistic energy-momentum relation for the free particle—in quantum mechanical form. Dirac soon discovered that there were no numerical coefficients for which his equation implied the Klein–Gordon equation. Instead of giving up the trans-equational analogy between classical magnitudes and quantum operators, Dirac introduced matrices—not, as Heisenberg, in place of the variables of mechanics, but as coefficients in his equation. The rest of the story is textbook stuff—matrices as coefficients required the wave function to quadruple itself. Nevertheless, each of the four components of a Dirac solution satisfied the Klein–Gordon equation, as required. And the introduction of matrices turned out to be a blessing, because now the phenomenon of electron spin—and of anti-matter—simply fell out of the equation.[64]

Recall, next, the Yang–Mills equations, which we mentioned in the context of a discussion of the role of symmetries in physics. But we could also mention it here, as a classical example of the use of a mathematical analogy in deriving an equation. The analogy was to electromagnetism. The idea was, as we noted, that their equations were to be related to isospin as the electrodynamical equations are to charge. It is true, of course, that isospin is related to charge; charge marks out the "direction" of isospin. Yet the Yang–Mills equations *ignore* the electromagnetic interaction with charge. When we consider the extremely salient fact that the primary analogy between electromagnetism and Yang–Mills theory was a speculative analogy between the abstract symmetries of the two kinds of field, we are forced to the conclusion that Pythagorean analogies guided Yang and Mills.

Our last examples of theory construction by mathematical ana-

---

[64] Namely, a solution of the Dirac equation consists of a quadruple of functions, which give, respectively, the probability (density) of finding, at a point of spacetime, an electron with spin up, a positron with spin down, an electron with spin down, a positron with spin up. We will return to Dirac's discovery later.

logies without physical ground derive from complex function theory, and involve the concept of an analytic continuation.

A function of a complex variable is called "analytic" in a region of the complex plane if it can be represented in that region by one or more convergent power series.[65] Equivalently, a function analytic in a region is one whose derivative exists in that region; this equivalence, which is highly nontrivial, holds only for complex functions, not for real functions, because the existence of the derivative of a complex function is a much stronger condition than the existence of the derivative in the real case.[66] This is because, in the complex case, the existence of the derivative at a point means that the differential quotient has not only to converge to a limit, but to converge to the *same* limit regardless of the direction from which we approach the point.

Now a complex power series, typically, does not converge everywhere in the complex plane (though it may), but within a certain "circle of convergence." (Where the power series converges everywhere, the radius of its circle of convergence is said to be infinity.) Thus, if a function is represented in a region by one or more power series, it is not immediately obvious that it can be extended to be an analytic function in wider regions. Remarkably, however, it can be shown that any such extension is unique—so that if we begin representing an analytic function by a power series in its circle of convergence, we may be able to "continue" the function analytically by the method of overlapping circles. And, fortunately, the historically useful functions turn out to be analytically continuable from the real line to the entire complex plane, with the exception of isolated singularities (such as zero in $\log(x)$).

The concept of analytic continuation is central in defining a concept which became important to physicists in the sixties—"crossing symmetry."[67] It was discovered that the equations of quantum field theory imply that the interactions among elementary particles are described, probabilistically of course, by analytic[68] functions of such parameters as the energy of the interaction. More: the equations imply that the functions describing interactions of particles have

---

[65] We have already discussed analytic functions in Chapter 2.

[66] It is even stronger than the existence of the nth derivative for every n.

[67] See Wightman 1969.

[68] More precisely, *piecewise* analytic, cf. Wightman, 120.

"crossing symmetry." That is, if a function, e.g., $f(E)$,[69] describes a collision of particles at energy $+E$, then if we analytically continue f to $-E$, $f(-E)$—which has no obvious physical meaning—nevertheless describes the expected results of a collision involving an antiparticle. For example, consider the function describing "Compton scattering," in which an electron hits a photon and both are affected. By applying crossing symmetry twice, we get the probability amplitude for the annihilation of an electron–positron pair to produce two gamma-rays (photons). Notice that the path that we pursue in extending the function from $+E$ to $-E$ need not stick to the real line; it can wander off into the imaginary realm in order to outflank singularities. Even on the real line, the function may be extended into regions which are "unphysical"—for example, where the energy of the system is less than its mass.[70]

So far, we have an absorbing mathematical property of quantum field theory. But now comes the induction: "all experimental evidence accumulated so far supports the view that crossing symmetry holds in Nature."[71] This prediction, made in 1969, was not merely a description of current equations; it was a constraint to be followed in constructing new theories, outside quantum field theory—just as "Lorentz invariance" is meant as a constraint on constructing new theories. Yet unlike Lorentz invariance, a mathematical constraint based on a physical idea (Relativity: equivalence of all inertial systems), the requirement of crossing symmetry as a feature of any physical theory of subatomic collisions is Pythagorean. For there is no way to express what all theories obeying this symmetry have in common, without using the mathematical analogy. To project that "all experimental evidence accumulated so far supports the view that crossing symmetry holds in Nature" is like saying, "all evidence so far supports the view that Presidents of the United States elected every 20 years are in trouble." Both are anti-naturalist analogies; the difference being that the former analogy is Pythagorean, while the latter analogy contains many other forms of anthropocentrism, forms that

[69] I am simplifying by leaving out variables other than energy. As Barry Simon has pointed out to me, it is quite important in crossing symmetry that one has more than one variable.

[70] Again, this point was impressed upon me by Simon.

[71] Wightman 1969, 124.

have played no recent role in science at all. This example shows how deeply Pythagorean analogies penetrated physical thinking in this century.

When a classificatory scheme becomes basic to scientific thinking, it can become invisible. Scientists "use" it without "mentioning" it.[72] The incongruity of a "naturalist" scientist making a Pythagorean prediction is not felt.

Readers who are experts in physics, however, may argue that crossing symmetry is not as Pythagorean as I claim. If we limit ourselves to quantum electrodynamics, the "electro-weak" theory, and any other relativistic theory which obeys the principles of "locality" and "causality,"[73] there is, in fact, a simple argument for crossing symmetry, which alleviates its Pythagorean character. Suppose that a collision occurs at a given point ("locality"). Then, in classical Special Relativity, none of the scattered particles can travel faster than the speed of light; otherwise, some observers would see the scattering before the collision, violating causality. In quantum theory, however, if a collision is localized, the Uncertainty Principle tells us that there is a finite probability of finding one of the particles outside of the "light cone," i.e., one that has traveled faster than the speed of light. Crossing symmetry resolves this contradiction. The behavior of a positron is exactly that of an electron moving backward in time. (And so, of course, for all anti-particles.) Thus the observer sees not particles moving backward in time, but anti-particles moving normally.

In cases such as these, the concept of analytic continuation plays little role. We have a general description into which we can substitute various situations; hence, it is not in any way mysterious that two state descriptions are analytic continuations of each other.

But there are theories (such as so-called "S-matrix theories") for which this argument does not apply. These theories do not attempt to describe the progress through spacetime of a "wave function" (or its generalization in field theory) from which predictions are extracted. Rather, the theories attempt to predict the results of scat-

---

[72] As Sidney Morgenbesser said, while giving W. V. Quine the gift of a necktie.

[73] Locality is the ability to localize an event at a point in spacetime; causality is the principle that the cause precedes its effects.

tering directly. In particular, no principle of locality is assumed. Furthermore, there is no general scheme into which we can "substitute" to yield scattering amplitudes for various kinds of events. To the contrary, crossing symmetry is a constraint built in to the theory on the basis of a Pythagorean mathematical analogy. For example, the present rage, "string theory," at least at present, violates locality—yet the theory is constructed to obey crossing symmetry.[74] Admittedly, string theory has its detractors, so the success of this particular analogy cannot yet be finally assessed.

The conception that the analytic continuation of an important function is also important is not limited to the phenomenon of crossing symmetry. For example, it has recently been discovered that if we "rotate" the complex plane 90 degrees, the basic formulas of quantum field theory are transformed into the basic formulas of statistical mechanics. This fact makes calculations possible which previously were impossible (instead of direct calculations in one theory, one "rotates" the problem, calculates in the other theory, then "rotates" back). It also has led to predictions that analytic continuations will lead to new discoveries and insight into old ones.[75] Since the concept of "analytic continuation" is, at present, a purely mathematical concept, any such prediction, even if based on past successes, is still a Pythagorean mathematical analogy.

* * *

The use of Pythagorean analogies is typical of twentieth-century physical thinking for two main reasons: the rich development of modern mathematics, and the lack of any alternatives. To conclude this section, then, I will review two venerable, endlessly-discussed tactics of scientific method, hoping to persuade you that—despite appearances—they are fundamentally Pythagorean, or at least that they contain an indispensable Pythagorean factor.

(5) A refuted law is used to test new laws—the "old" law is stipulated to be a special or limiting case of any "new" law.

---

[74] I am not claiming that crossing symmetry is the major idea in string theory; merely that it is one constraint on string theory, one which at present has no physical basis.

[75] See Manin 1981, 80–4.

(6) A refuted law—false by definition—is nevertheless used to derive new laws.

Strategies (5) and (6) presuppose that mathematical structures are more robust than the laws that instantiate them. Even if a law is refuted, its mathematical form (symmetry) must play a role, one projects, in future developments. Such a projection is Pythagorean, because we cannot characterize physically what we mean by "mathematics."

Consider (5). An obvious, naturalistic explanation of the scientists' desire to have new laws yield the old laws is that if law A is a special case of law B, then all the data that was thought to confirm law A are transformed automatically into data confirming law B. Even if, as usual, law A is only a limiting case of law B, so that technically the laws are incompatible, we can treat the evidence that was thought to confirm A as approximate, to within a given experimental error. In that form, the evidence *can* confirm law B, since law B predicts the experimental error of law A. Thus, we can say, that when law B yields law A as mathematical limit, the confirmation of A is transferred to B. I shall call this phenomenon "evidence transfer."

But if we treat the evidence for law A as merely approximate, there is no need to have law A, in its exact mathematical form, as a limiting case of law B. All we need is that law A imply that *law B is true to within experimental error*, a different matter entirely.[76] Hence evidence transfer does not entirely explain why scientists prefer that new laws be limiting cases of the old.

In fact, there are philosophers, like Feyerabend (Feyerabend 1978), who condemn the whole idea of anchoring "law A" in advance as a criterion for any new law. According to Feyerabend, scientists should actively seek new laws that do *not* yield the old laws, but rather explain why we *thought* that the old laws were confirmed by the evidence. This is the antidote to dogmatism, the enemy of true science.

Another reason, though no more palatable to Feyerabend, why scientists prefer to have old laws be the limit of new laws is—inertia. Scientists invest an enormous amount of energy and time in learning

---

[76] This point has been made many times by "scientific realists"; I'm using it for my own purposes.

a physical theory. Learning a physical theory means, above all, learning the "mathematical methods" appropriate for the theory. This includes tricks for solving or approximating solutions for equations, and many other things. If a new theory yields, mathematically, the old, then all of those mathematical techniques are still useful. An extremely important instance of this is the use of eigenvectors in classical, and then in quantum, mechanics.

Here too, and even more so, Feyerabend's criticism is apt: isn't it just dogmatism (and certainly anthropocentric) to assume that *nature* cares how much time we have invested learning mathematical methods? Isn't yielding to inertia an intellectual sin?

Actually, I think that the preference of scientists for laws that yield the old laws, mathematically, is not just inertia. Physicists take the mathematical passage from law B to law A as striking evidence of a discovery:

> It is possible to know when you are right way ahead of checking all the consequences. You can recognize truth by its beauty and simplicity. It is always easy when you have made a guess, and done two or three little calculations to make sure that it is not obviously wrong, to know that it is right . . . . We have to find a new view of the world that has to agree with everything that is known, but disagree in its predictions somewhere . . . . If you can find any other view of the world which agrees over the entire range where things have already been observed, but disagrees somewhere else, you have made a great discovery. (Feynman 1967, 171)

It is not inertia operating here, pace Feyerabend, but Pythagoreanism. It is the mathematical "beauty and simplicity" which we encounter in having a new theory, based on a "new view of the world," nevertheless yielding the old theory as a mathematical theorem. That is, Feynman asserts that it is very significant that two theories that are conceptually different are mathematically related, and it is this relation that signals a "great discovery."

But I foresee an objection: scientists have come to expect that successful new laws yield the laws they replace. Quantum electrodynamics yields Maxwell's "classical" electrodynamics at the macroscopic level. Maxwell's theory yields Fresnel's wave theory of light under the conditions appropriate for the emission of electromagnetic radiation, and of the appropriate frequency. Fresnel's wave theory goes over into geometric optics (Snell's law), where effects such as diffraction are negligible. There are two further approximations: geometric optics yields linear optics in the limit of small angles of incidence; linear optics yields so-called "Gaussian" optics, where the system exhibits rotational symmetry (as in lenses).[77] Thus, when a new law yields the law it replaces, it fits into a historical pattern which could provide additional evidence for the law, beyond evidence transfer. No Pythagoreanism here, just historical evidence.

But this use of historical evidence itself presupposes Pythagorean thinking. For what connects the various case histories is the higher-order mathematical analogy: one law yielded another one *mathematically*. There is nothing beyond the mathematical analogy itself to ground the historical evidence if, as I argue, there is no naturalist definition of "mathematics."

Given the Pythagorean point of view, however, Feynman's procedure makes perfect sense.

A major function of a law, for the Pythagorean, is to pick out a mathematical structure, or symmetry, that can be used to describe nature.[78] Furthermore, the structures the law picks out may and often do survive the death of the physical ideas that motivated the law or even the law itself. For example, Einstein laid down the constancy of the velocity of light (in a vacuum) in all inertial frames as a postulate of physics, giving what amounted to verificationist arguments to support this. From this it followed that an event with space-time coordinates in one inertial frame would have, in another inertial frame, not only different space coordinates (this was obvious), but also clock a different time. (Otherwise the speed of light will go down relative to an inertial frame speeding in the same direction as the light.) The formula linking the coordinates (x,y,z,t) of an event in one frame with those (x',y',z',t') of the *same* event in another, in order that the speed of light should remain constant, is called a Lorentz transformation. Now Einstein argued that every law of nature must be invariant under a Lorentz transformation: a law of nature $f(x,y,z,t) = 0$ must obey the condition

---

[77] This hierarchy is discussed in Guillemin and Sternberg 1990a, ch. 1.

[78] As we have seen, the expression "symmetry instantiated by nature" is very misleading.

$$\forall xyzt: f(x,y,z,t) \leftrightarrow f(x',y',z',t').$$

For, if not, one could tell which inertial frame one was in, i.e., how fast one was moving in absolute space, by seeing what laws hold, a concept Einstein rejected.

The Lorentz symmetry of equations has been accepted as a permanent feature of physics, whereas the physical conception that allowed Einstein to write down the Lorentz transformations themselves is quite dispensable. As Levy-Leblonde points out, although all evidence is consistent with the current presumption that photons have mass zero, if better measurements accorded photons a little mass, their speed would no longer be c (Levy-Leblonde 1979). This sounds like an absurdity but is not:[79] the constant c has long functioned in physics as a "conversion constant" which allows time to be treated like space.[80] That there actually be an object that travels with velocity c is not necessary or even relevant.

The sentiment that mathematical structures outlive the physical conceptions that embody them is well expressed by Steven Weinberg:

> . . . very often beautiful mathematics survives in physics even when, with the passage of time, the principles under which it was developed turn out not to be correct ones. For example, Dirac's great work on the theory of the electron was an attempt to unify quantum mechanics and special relativity by giving a relativistic generalization of the Schrödinger wave equation . . . the principles that Dirac was following have been aban-

---

[79] For those familiar with Kripke 1980, it may be illuminating to regard 'c' as a "rigid designator." The definite description "the speed of light" was never supposed to be a definition of, but to fix the reference for, 'c'. If photons have mass, then there is no single speed of light. Semantically, then, 'c' has no reference. Nevertheless, we often give the *speaker* the benefit of the doubt when he fixes the reference of a term mistakenly using a definite description that has no reference in the actual world. We correct the definite description for the speaker and assign the reference of the corrected description to the speaker's term. In Donellan's example, "The man over there drinking a martini," when he is in fact drinking vodka, we assume the speaker meant the man drinking the vodka. Similarly here: even if it should turn out that photons have mass, so there is in fact no one speed of light, Einstein's reference was to a certain constant c, which he *thought* was the speed of light, and which *would* have been the speed of massless photons. We can plausibly assign the *speaker's* reference to Einstein's 'c' based on this corrected description. The nomenclature "semantic reference" and "speaker's reference" is also due to Kripke.

[80] The quantity ict is what is on a par with the three space dimensions.

doned, but his beautiful equation has become part of the stock and trade of every physicist; it survives and will survive forever. (Feynman and Weinberg 1987, 110)

A more blatant Pythagoreanism would be hard to find.

Let's look more closely at how strategy (5) preserves mathematical structure. Suppose that law A (an equation, say) turns out to be a limiting case of law B. Then, typically, symmetries of law B become symmetries of law A. But this means that, in the typical case, law A has more—not fewer—symmetries than the law which replaces it. Now, on the one hand, this makes physical sense: a world in which undifferentiated primal matter filled an infinite space, with nothing happening, would be extremely symmetrical.[81] Once forces start operating, the pastoral symmetry is simply broken. But if law A has more symmetry than law B, symmetries are not being preserved; they are being lost.

The solution to this riddle is subtle. Consider a sphere being shrunk to a point. The sphere has rotational symmetry, and so does the point, its limit. Nevertheless, there is an obvious sense in which the sphere has "more" symmetry than the point, in that the rotational symmetry has become degenerate in the limit, on account of the loss of two dimensions. Something like this happens in the interesting cases. As physics progresses, the symmetries tend to become higher-dimensional, as we postulate more and more dimensions[82] for the laws to govern.[83] To put it another way, new laws can unfold greater

[81] There is, therefore, a difference between symmetry and order. We perceive as ordered a universe with an optimum degree of symmetry, not a maximum degree of symmetry. This probably means that the concept of order is anthropocentric.

[82] The term "dimension" need not mean here physical dimensions. For example, in a system with two particles, we can think of the system as having six abstract dimensions, since each particle moves in three. Any degree of freedom that satisfies the appropriate topological condition can be called a dimension. That said, it should be noted that physicists are speculating that the four dimensions of space and time of our physical universe are a small number of those that exist.

[83] Thus the hierarchy of successively complex optical theories corresponds to successively more complex symmetry groups. See again the first chapter of Guillemin and Sternberg 1990a, where the authors go so far as to *identify* the various theories with the symmetry groups by the following isomorphism: each optical system described by a given theory is associated with a member of a corresponding group, and the physical operation of composing the systems is associated with the group multiplication.

symmetry, because they enlarge the very arena in which the laws are played out. There is simply more to *be* symmetrical.[84]

Strategy (6), a sort of converse of strategy (5), occupies a place of honor in the physicist's arsenal. It is the strategy applied by Newton in deriving the inverse square law from Kepler's laws, as part of his program "from the phenomena of motions to investigate the forces of nature, and then from these forces to demonstrate the other phenomena."[85] Newton imagines a Keplerian mass point in an elliptical orbit, around an immovable center of force, located at a focus.[86] A mathematical argument shows that such a point always accelerates toward the focus of the ellipse in inverse proportion to the square of the distance. Since force is proportional to acceleration, we are done.

Pierre Duhem has a famous criticism of this argument, in which he denies that Newton could have deduced the inverse square law from the experimental data.[87] For Newton, as is well known, raised the inverse square law to the status of a universal phenomenon: every body attracts every other one by the same force that keeps the planets in their orbits. Thus, Duhem argues, the planets must attract one another, interfering with their Keplerian motion. But then their orbits do not imply the inverse square law. Another problem is the third law of motion. If the sun attracts the planets, then the planets attract the sun. Hence we cannot speak of a fixed center of force, and the derivation is unsound. Thus Newton made use of data which, according to his own theory, must be false.[88] Hence the data cannot be used as true premises in a derivation, and Newton misrepresented his own procedure. It would have been more honest, Duhem felt, for Newton to do the opposite: to assume the inverse square law for the purpose of deriving its consequences and then just check the consequences (the "hypothetico-deductive method" favored by Duhem). Thus, for example, the inverse square law applied universally implies

---

[84] Thanks to Harry Furstenberg and Joel Gersten for helping me with this point.

[85] Newton 1934, Preface to 1st edn.

[86] Kepler's first two laws are sufficient for the demonstration: that planets revolve in an elliptical orbit with the sun at a focus; and that a radius vector from the sun to the planet sweeps out equal areas in equal times.

[87] See Duhem 1962.

[88] The structure of this argument is: if p is true, then it is false; hence p is, in fact, false.

that the Kepler laws are approximately true even when we take into account the motion of the sun and the interplanetary perturbations.

Duhem's criticism is cogent, but Newton's argument can be reconstructed as follows. As a rule of thumb, consider Principle (N):

(N) Whenever a physical phenomenon is approximately describable by a mathematical structure S, we assume that there is something about the phenomenon that is exactly describable by S, and the deviation from S is caused by a perturbation (i.e., we don't *just* say that the description by S is inexact).

Kepler's laws approximately represent the motions of the planets; i.e., they are correct to within experimental error (in the time of Newton), particularly in the case of the planets close to the sun. We therefore assume that the symmetry addressed by Kepler's laws is the symmetry of something about the planets. To discover what that is, we study the model of a single planet orbiting around an immovable sun (located in the focus of the ellipse), and for this case, there is a mathematical proof that the sun is exerting an inverse square force to keep the planet in orbit. We then go on to study this force—which Newton discovered was none other than the force of gravity—rather than Kepler's laws. In the particular case of planetary motion, Newton was very lucky, since the actual deviation from Kepler's laws observed by later astronomers was completely accounted for by gravitation. That is, even the perturbing force on the planets, causing the deviation from Kepler's laws, was the same force Newton had derived using his ideal model.

Note, however, that Principle (N) is deeply Pythagorean, since, once again, we cannot independently specify what a "mathematical structure" is. Chess, for example, is not a mathematical structure, for reasons that are ultimately subjective. Thus we must, it appears, attribute Pythagoreanism to Newton to make sense of his procedure. Principle (N) essentially says, as a rule of discovery of course, that every approximate symmetry is to be treated as a *broken* symmetry.

Are there any other ways to rationalize Newton's procedure, ways that do *not* assume Pythagoreanism? Consider one of the deepest studies of Newton's procedure, Harper 1990. Harper points out that there is a sense in which Kepler's orbits continue to exist even though they are perturbed. Namely, by the law of superposition (i.e., the linearity of the homogeneous equations of motion), the actual orbit of each planet is the vector sum of all the motions due to all the forces acting on the planet. The dominant component is due to the sun and is an ellipse. (I am simplifying, of course, by assuming the sun to be immovable for the sake of argument.) We can then view Newton's procedure as straight *deduction* from Kepler's laws, applied not to the actual motion of the planets, but to a component.

Harper's analysis is certainly correct, but it presupposes mine; i.e., it does not replace Pythagoreanism, but adds to it. In corroborating the inverse square law by Kepler's "observations,"[89] Newton had no basis for assuming that the Kepler motions were *true of* even components of planetary motion, if he could not distinguish between the "real" symmetry of the motions and the perturbation. Only Principle (N) does that. *Given* Principle (N), it is fine to analyze the situation as Harper does, though Principle (N) can be used also in cases where the laws of motion are not linear, and where, therefore, the mathematical structure does not even describe a component of the actual motion. Hence Principle (N) is deeper than Harper's principles.

I have argued that Principle (N) affords a "rational reconstruction" of Newton's reasoning which avoids Duhem's censure. That is, even if Newton did not rely on Principle (N) consciously, his actual behavior is Pythagorean. But in fact, Newton made declarations which evoke Principle (N). Principle (N) gives a very neat interpretation of Newton's fourth "rule for reasoning in philosophy" (*Principia*, Book III):

> In experimental philosophy we are to look upon propositions inferred by general induction for phenomena as accurately or very nearly true . . . till such time as other phenomena occur by which they may either be made more accurate or liable to exceptions.

Little meaning can be given to "very nearly true" here without Principle (N). Once we are confirming the approximate truth, rather than the truth, of a proposition, there is no end to the propositions

---

[89] Newton never gave Kepler credit for having discovered "laws." This is mean-spirited, but also reflects Newton's belief that physical laws are those which specify the forces of Nature, i.e., dynamical, not kinematical, laws.

we are confirming. With Principle (N), of course, the weather clears: propositions are mathematical propositions or structures. These don't grow on trees. For example, the inverse square is a mathematical structure, but the inverse of 2.00000000093 is not.

* * *

The "Newtonian" strategy is a very powerful method of discovery in contemporary physics. We have already discussed the so-called "missing" lines of the spectrum of hydrogen. Missing lines mean that there are transitions that the hydrogen atom, for some reason, does not undergo. An explanation for this could be the existence of a hitherto undiscovered symmetry of the system, one which imposes a "new" law of conservation, preventing the "missing" transitions. An example is the symmetry known as parity.

But, in fact, the transitions, the spectral lines, are not missing. They are simply fainter than the others. According to Pythagorean strategy (N), we are to regard the fainter lines as perturbations of what is a broken symmetry, where "symmetry" is defined by mathematicians.

What is more, the effect that breaks the symmetry may itself be governed by a symmetry. (In the case of Newtonian gravitation, what broke the gravitational symmetry of Kepler's laws is itself a gravitational perturbation.) Here is another example of Principle (N), and thus Strategy (6), which illustrates this idea—that what breaks the symmetry may itself be governed by a symmetry.

A lambda hyperon (Λ^0) is a hadron with isospin zero and charge zero. We find that a hyperon "weakly" (i.e., slowly) decays into a proton (p) and negative pion π^-, or a neutron (n) and a neutral pion π^0:

$$\Lambda^0 \to p + \pi^-$$
$$\Lambda^0 \to n + \pi^0.$$

The first type of "weak" decay occurs *approximately* ⅔ of the time; the second, about ⅓ of the time. And both reactions *fail* to conserve isospin: a proton has isospin ½; a neutron, $-½$; a neutral pion, 0; a negative pion, -1.

Here physicists guessed: if the statistics (i.e., ⅔ to ⅓) were exact, we could calculate as follows. Multiply the two-dimensional nucleon space by the three-dimensional pion space, in the sense of tensor product (cf. above). We then get a six-dimensional space. Within this space, the neutron is *formally* represented as the following linear combination:[90]

$$n = \sqrt{1/3}\, \pi^0 \otimes n - \sqrt{2/3}\, \pi^- \otimes p.$$

Squaring the coefficients, as we do in quantum mechanics, we get the *exact* probabilities $2/3$ and $1/3$ we are looking for. Physicists thus conjectured (using Principle (N)) that there is some natural, if weak, interaction, which increases the isospin from zero to $1/2$,[91] so that it can decay as the above. Though the latest theories of matter give insight into the process, these allow other possible scenarios for the hyperon decay, and it is not yet clear what rules them out. Nevertheless, the Pythagorean conjecture, that the weak decay of the lambda hyperon proceeds via an SU(2) mechanism, seems vindicated.

A saying popular in the sixties was: "To be a particle physicist, you need two things: familiarity with the Greek alphabet, and an ability to recognize fractions."[92] This is a beautiful definition of a third-century Pythagorean.

---

[90] Cf. Sternberg 1994, 213 ff. for the details of this group theoretical calculation.

[91] It is not that the hyperon actually becomes a physical neutron before decaying—it is abstractly represented as one before decaying.

[92] Shlomo Sternberg told me this.

5

Formalisms and Formalist Reasoning in Quantum Mechanics

In quantum mechanics, formalist analogies often take the form of pseudodeductions: instead of preserving truth, formalist reasoning establishes meaning. Formalist reasoning shows that the extension of the formalism to new situations is constrained by the formalism itself. Such reasoning is like what my student and colleague Meir Buzaglo calls "strongly forced extensions" in the history of mathematics. For example, when mathematicians extended the formalism of "raising to a power" to cover zero, negative, rational, real, and even imaginary powers, they found that they had little or no choice in defining these concepts, so long as the essential syntax of raising to a power was preserved.[1]

But there is a major difference, too: physics, unlike mathematics, is subject to empirical tests. That a physical formalism can be extended only if certain empirical conditions hold is not a naturalistically valid reason that these conditions in fact hold (who says that formalism should be so extendable?). In fact, I maintain, much physical research in the present century has been, in the first instance, inquiry into our own formalisms, and only secondarily into nature.

---

[1] For example, we want to preserve rules like

$$x^{\alpha}x^{\beta} = x^{\alpha+\beta}$$
$$\frac{x^{\alpha}}{x^{\beta}} = x^{\alpha-\beta},$$

but are willing to sacrifice $x^2>0$. Buzaglo defines the notion of "forced" extensions and "strongly forced" extension, using the techniques of model theory.

The scientist investigates a formalism to "see what it means," as though it were the Handwriting on the Wall in the Book of Daniel. Therefore, much research is based on anthropocentric premises. And the *success* of such research makes the universe *look* anthropocentric. (I do not say that the success of anthropocentric research confirms one or another anthropocentric premise, because dogmatic naturalists can always rule out the confirmability of anthropocentrism. Still, I can say that even for dogmatic naturalists, the success of anthropocentric research is a *mystery*, and that the usual strategies of explaining away anthropocentrism do not obviously work.)

In Appendix A, I "derive," by formal means, such results as the Heisenberg Uncertainty Principle, the quantization of angular momentum, and some of the properties of electron spin. Of course, the results themselves were known before anybody thought of showing how these results are latent in the formalism of quantum mechanics. My main object there is simply to show that the Hilbert space formalism is descriptively applicable to quantum mechanics, and to point out that no explanation for this is *presently* known.

Nevertheless, the formal derivations of empirical results are germane to the context of discovery also. These formal derivations take the form of showing that there are severe restrictions on how the formalism can be extended—for example, we must either quantize angular momentum, or give up the formalism. This suggests to the scientist that this is the route to further discoveries, that one must "do the same thing as before" to make these discoveries. In other words, the motivation for this sort of research is—an analogy.

But we have already seen the slippery nature of "doing the same thing as before." If the analogy in question is illegitimate, then, it is simply not true that we *are* "doing the same thing as before." For the naturalist, in particular, using the formalism to make discoveries by analogy to other cases of extension, is, in my opinion, ruled out; the naturalist is blocked from a claim based on "doing the same thing again."

This chapter shows how the quantum mechanical formalism was actually used to make a prediction in this way—how scientists assumed that the formalism continues to "track" nature even when it is altered by extension.

Our story has to do with the extension of the quantum mechani-

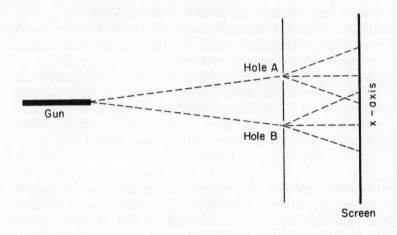

FIGURE 2

cal formalism to a two particle situation. Let us first recall a fundamental property of the quantum mechanical formalism—one which
I will not attempt to derive at all. This is the Principle of
Superposition. It says that if vector Ψ_A describes a system in state A,
and vector Ψ_B describes a system in state B, and we have no way of
deciding, even with probability, whether the system is in A or B, then
the vector describing the situation that the system is either in state A
or in state B is (proportional[2] to) $\Psi_A + \Psi_B$.[3] A familiar example of
this is the "two hole" experiment (see Figure 2). We shoot one particle at a wall with two tiny holes, in back of which a screen scintil-
lates wherever a particle lands. We can take the screen as the x-axis.
Suppose that, with hole A open, the vector that gives the position
data of the particle is Ψ_A; then the probability that it will strike at a
point x is $|\Psi_A(x)|^2$. With hole A closed and B open, suppose the prob-
ability curve is given by $|\Psi_B(x)|^2$. (Both curves might look like bell

[2] The sum of the vectors might have to be multiplied by a constant, in order that
the probabilities still add up to one. This is called "normalization." I will adopt the
attitude that the Principle of Superposition is simply part of the rules of the formalism.

[3] The ground of this principle is still controversial. For a readable account of the
principle, see Albert 1992.

curves, each with the maximum opposite the corresponding hole.) Then, with both holes open, and the experiment conducted so that there is no way to detect which of the two holes the particle traverses, the probability curve is $|\Psi_A(x) + \Psi_B(x)|^2$, which is *not* the sum $|\Psi_A(x)|^2 + |\Psi_B(x)|^2$ of the two probability curves.[4] On the contrary, the two vectors can interfere with each other, and the screen will have areas forbidden to the particle. (If a whole beam of particles is shot, the screen will contain dark areas.) The reason is that the coordinates of the vectors are complex numbers that have "directions" as well as "length."

We confront here one of the famous "paradoxes" of quantum mechanics: how can opening *another* hole in the wall *restrict* the movements of the particle? One would think the opposite is the case: the particle now has more degrees of freedom. Much ink has been spilled on this and other "paradoxes" of quantum mechanics[5] and the matter is in good hands. This book is not about the mysteries of quantum mechanics, but the mystery of its discovery; and, as here, the mystery of its extension.

Let us see if we can extend the formalism to deal with *two* particles, particles 1 and 2. Each position "axis" will now represent the position of two particles, not one. Indeed, if we consider position in three-dimensional space, then each position axis will correspond to a value for the following six variables: $x_1, y_1, z_1, x_2, y_2, z_2$, the first three numbers giving the position of particle 1; the second three, that of particle 2. The coordinate of the state vector will then be a (complex-valued) function of six variables: $x_1, y_1, z_1, x_2, y_2, z_2$. The absolute square of this function then gives the probability that particle 1 is at the place (x_1, y_1, z_1) and particle 2 is at (x_2, y_2, z_2). The six-dimensional space defined by the variables $x_1, y_1, z_1, x_2, y_2, z_2$—the "direct sum" of two copies of three-dimensional Euclidean space—is called the "configuration space" of the system of two particles, the space on which the function $\Psi(x_1, y_1, z_1, x_2, y_2, z_2)$ is defined. It is important not to be confused about the difference between the configuration space, where the system resides, and the linear space, where the state vector resides.

[4] Actually, it is $\frac{1}{2}|\Psi_A(x) + \Psi_B(x)|^2$; otherwise, the sum of all the probabilities would be 2.

[5] I refer the reader once again to Albert 1992.

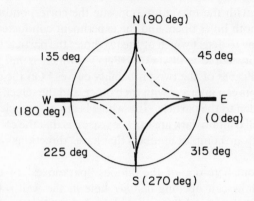

FIGURE 3

After Wilczek 1991, let us consider the following experiment (see Figure 3). Two particles with equal charge and mass are "shot" simultaneously at each other at equal velocities from guns that are located an equal distance east and west, respectively, of the origin. The particles are scattered in different directions, but the difference between the two directions must be exactly 180°; it is sufficient to know the direction of either particle, to know the direction of the other one[6] (the convention will be that the direction given is always the angle of the western particle, the particle that comes from the west). Thus, it is possible, but not necessary, that one particle is scattered due north, and the other, due south. In fact, either particle could go due north, or 90°. In theory, of course, we need three coordinates to locate a particle (x,y,z, or, in spherical coordinates, r,ϕ,θ, where θ is always the "latitude" and ϕ the "longitude"). But suppose we care only about the angle ϕ—the "scattering angle" (with respect to a horizontal plane). We can then speak of a function $\Psi(\phi_1,\phi_2)$ of two variables only, the amplitude for particle 1 to end up at ϕ_1 at the same time as particle 2 ends up at ϕ_2. (N.B.: we will not be able to ignore the three-dimensionality of space for long, though.)

[6] This fact is not special to quantum mechanics, but follows from the classical principle of conservation of momentum. I have simply set up the experiment, following Wilczek, in such a manner that the outcomes are easy to calculate.

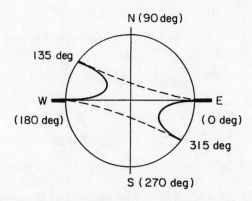

N (90 deg)

135 deg

W
(180 deg)

E

(0 deg)

315 deg

S (270 deg)

FIGURE 4

Two ways of arriving at the same result: Particle 1 incoming from west; Particle 2 from east. One particle arrives at 135°; the other at 315°.

As in the two-hole experiment, there are always two ways to get *one* particle at any given place (i.e., "longitude" f). For example (see Figure 4), a particle could end up at 135°, and the other, perforce, at 315°, if either:

    (A) particle 1 (incoming from west to east) is deflected backwards at an angle of 45° from its incoming path,

or:

    (B) particle 2 is deflected forward at an angle of 135° from its incoming path.

If the particles are nonidentical, though having the same mass and charge—in other words, if there is *any* property that distinguishes the two particles—then we can distinguish between cases (A) and (B). In that case, the Principle of Superposition does not apply. So to get the probabilities to have *one* particle landing up at angle ϕ (and, of course, we cannot have them both ending up at the same place), we simply add the probabilities of case A and that of case B, arriving at

$$(|\Psi(135°,315°)|^2 + |\Psi(315°,135°)|^2).$$

If, however, the particles are identical, we would expect (given the Principle of Superposition) the following result:

$$|\Psi(135°,315°)| + |\Psi(315°,135°)|^2.$$

In general, we can write the following expression for the probability that exactly one particle will be deflected in the direction ϕ_1:

(1) $$|(\Psi_{nonid}(\phi_1,\phi_2) + \Psi_{nonid}(\phi_2,\phi_1))|^2$$

Let's consider the special case where one particle is deflected due north, the other due south. By symmetry, clearly

$$\Psi_{nonid}(90°,270°) = \Psi_{nonid}(270°,90°).$$

In other words, in the case of nonidentical particles, we get the same amplitude (and the same probability) for either particle to go north.

In this case the calculation is easy, and similar to the one for the two-hole experiment. The probability for one particle to be deflected due north, in the case of nonidentical particles, is

$$(|\Psi_{nonid}(90°,270°)|^2 + |\Psi_{nonid}(270°,90°)|^2) = 2|\Psi_{nonid}(90°,270°)|^2,$$

and in the case of identical particles

$$|\Psi_{nonid}(90°,270°) + \Psi_{nonid}(270°,90°)|^2 = 4|\Psi_{nonid}(90°,270°)|^2,$$

i.e., twice as likely—just as before. And, in fact, this is exactly what happens—sometimes. For *some* pairs of identical particles, that is, our calculation is exactly what the Principle of Superposition predicts.[7]

Consider the case of two *electrons* shot from the guns of our experiment. What is the probability that an electron will be deflected due north? The shocking experimental answer—if we are still capable of shock after the Principle of Superposition—is: zero. How is this possible? And could there be any other behavior then the two just described: either twice as much probability at due north or zero? (The latter kind of identical particles are called "fermions," and the former "bosons," in memory of the great physicists, Fermi and Bose, who discovered them.)

[7] And particles here can mean also: nuclei, i.e., clusters of particles that can be treated as a unit for our purposes. If you would like an example of "elementary" particles described by this calculation, the pions will do. Yet for other pairs of identical particles, there is the most startling result.

The explanation of this amazing phenomenon (interference, as it were, yet different from the interference of the two holes above) is typical of twentieth-century physics. No new laws will be put forward. Rather we will return to our formalism, and seek there information that we never noticed before—information conveyed nondeductively. This information will exhibit new possibilities in our formalism, and thus new possibilities in nature.

Consider the case of identical particles once again. So far we have discussed probabilities, but really we need a "state vector" to describe a system in quantum mechanics. What kind of linear space will we have? (The linear space is defined *on* the configuration space, since each point of the configuration space is considered an "axis" of the linear space.) What is the coordinate function, the psi-function, of the system? For nonidentical particles, we can get a function of two positions (six coordinates), $\Psi(x_1,x_2)$, where bold letters mean vectors, i.e., ordered triples. (In our experiment with the two guns, we ignored all coordinates except for the two "longitude" angles. As we shall see presently, the missing coordinates play a role even if we ignore them.) The space of six coordinates is called the "configuration space." The points of the configuration space, in turn, form the "axes" of our linear space, where the state vector resides.

In the case of identical particles, however, we cannot—by definition—tell the difference between the situation described by (x_1,x_2) and that described by (x_2,x_1)—i.e., particle 1 at place x_1 and particle 2 at place x_2—or the reverse. (This is why we didn't just add the probabilities in the first place.) Each of these describes the same situation, one particle at each place. What this suggests is that we *identify* the location[8] (x_1,x_2) with the point (x_2,x_1). This is like taking a piece of paper and folding it along its diagonal. Indeed, in our gun experiment, where we had only two coordinates, the angle of the first particle and the angle of the second, we get a two-dimensional graph, and we can visualize the folding quite easily. In other words, the psi-function should be thought of as a function on "half" the number of points when the particles are identical. A better way to put the matter is that we consider the pair $<(x_1,x_2),(x_2,x_1)>$ to be a member of

---

[8] A reminder: the "locations" here are not in ordinary three-space, but in the configuration space, which is six-space, $E^3 \oplus E^3$.

an equivalence class, and the psi function is now thought of as a function on pairs of equivalence classes. We might write $\Psi_{id}(X_1,X_2)$, with capital letters, to express this idea. The capital letter variables X_1, X_2 range over the equivalence classes of locations. The subscript "*id*" means that we are defining a new state vector, one that expresses maximum information concerning a system of two *identical* particles.

What are the pairs (x,x)—known as the *diagonal*—equivalent to? Only to themselves, thus there would have to be some equivalence classes with only one pair of positions. The way out here is simply to omit these points (take out the diagonal). I emphasize that removing the diagonal makes no difference,[9] because if the psi-function is continuous, then the values of the psi-function for points near the diagonal will determine the values on the diagonal. In other words, the value of $\Psi_{id}(X_1,X_2)$, where $X_1 \approx X_2$, would have to determine the meaning (the value) of $\Psi_{id}(X_1,X_1)$ or $\Psi_{id}(X_2,X_2)$. And, by the way, in the experiment now under consideration, the physics of the situation guarantees that the two particles can *never* land at exactly the same place, since each particle is distant 180° from the other. I stress this because the reader is quite likely to suspect "hanky-panky" in what follows.

Let's look at what happens when we take a six-dimensional (configuration) space composed of (the "direct sum of") two copies of ordinary three-dimensional Euclidean space: remove the pairs $<x,x>$ (the remainder will still be a six-dimensional space), and then "identify" every point $<x_1,x_2>$ with $<x_2,x_1>$. Since this still gives a six-dimensional space (we just "folded it over"), it is hard for us to visualize. To make things easier, indeed to reduce the problem to a three-dimensional one, consider this space, not from the point of view of the Cartesian coordinate system, but from the reference frame of one of the two identical particles. Each particle thinks it is the center of the universe, stationary there—and the other particle

[9] The diagonal of a sheet of paper is a line, but the diagonal of the six-dimensional space (the three coordinates of each particle) in which we locate two particles is an entire three-dimensional Euclidean space. In general, the diagonal has half the dimensions of the whole configuration space.

Only in two dimensions does the removal of the diagonal make a real difference—it divides the plane into two separate pieces; the plane is no longer a connected set. We will have occasion soon to remark upon the special properties of two dimensional manifolds.

moves around in a three-dimensional space. Note that the "diagonal" has now become a single point—the center of the universe, the origin, since only the other particle can move—and to say that the particles are on the diagonal is to say that they are at the same point in space.

So we take ordinary three–dimensional space, and puncture it only at the point (0,0,0), where one of the particles stays fixed, and the other particle never goes (or at least, we ignore it, if it goes there). From this point of view, what does it mean to replace a point $<x_1,x_2>$ with $<x_2,x_1>$—i.e., for the particles to switch positions? It means (from the parochial point of view of our "geocentric" particle, which thinks it never moves) for the *other* particle to reflect itself through the origin to the diametrically opposite position. That is, it moves from (x,y,z) to $(-x,-y,-z)$: a turn of 180°, if you prefer.

To make things even simpler, let us consider, not the whole space (x,y,z), but only those points within some radius r from the origin— a sphere and its interior (with the center removed). In Figure 5, each point in the "northern hemisphere" is identified with its diametrically opposite point in the "southern hemisphere." Let's call the result the "Punctured Projective Sphere (PPS)." This is a space that we cannot really visualize, but we can try to interpret the projective space in terms of our own. Figure 5, then, is a picture of the Euclidean sphere,[10] in terms of which we will try to grasp the unpicturable PPS.

For example, a closed loop in the PPS corresponds to a *pair* of closed loops in the Euclidean sphere, one the reflection of the other. In Figure 5, for example, the two loops are supposed to lie in planes parallel to the equatorial plane—one loop in the northern hemisphere, the other in the southern hemisphere—and points a and b correspond, respectively, to points a' and b'. Notice that we can shrink both loops simultaneously to diametrically opposed points while remaining in the Euclidean sphere; thus, the *one* loop in PPS to which the two correspond, can be shrunk to a single point—the point to which the two points in the Euclidean sphere correspond.

Now move the two loops continuously toward the equatorial plane. When they arrive there, the two loops coalesce into two simul-

---

[10] By "sphere" I mean to include the interior as well—what mathematicians sometimes call a "ball."

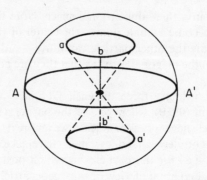

FIGURE 5. The Punctured Projective Sphere.
The loop in the northern hemisphere is equivalent to its mirror image in the southern hemisphere.

taneous trips around the missing center (from different starting points, however). We can also reverse the process by a continuous motion and bring back the two loops. Thus, the loop in PPS that corresponds to the two trips around the equator can also be shrunk to a point.

Suppose, now, we take two simultaneous trips around *half* the equator—from A to A' and from A' to A. The "trip" to which this corresponds in PPS is from the *pair* <A,A'> to the *pair* <A',A>. But these pairs are diametrically opposed to each other. Hence the pairs <A,A'> and <A',A>, though different on the Euclidean sphere, are one and the same on PPS. In other words, by taking two simultaneous trips around only *half* the equator, we have arrived back where we started from. (Not on the Euclidean sphere, remember, but in our unimaginable PPS.) That is, the pair of trips that we took, correspond, in PPS, to a closed loop. Yet it is not hard to see that we cannot shrink this closed loop to a point: if we move the two half equators off the equatorial plane, we will get two half loops in the northern and southern hemisphere. Thus, only on the equatorial plane does this special "PPS-loop" stay a loop. But the equatorial plane is punctured. To get the loop off the equatorial plane (i.e., to turn the loop into two small Euclidean loops), we have to complete (simultaneously, twice) the entire equator.

In sum, in the Punctured Projective Sphere, though there are loops that cannot be shrunk down to a point, any double loop can. (This fact can be given a mathematical proof in algebraic topology.)

Let us abandon, now, the point of view of one of the particles, and treat of both particles with one function $\Psi_{id}(X_1, X_2)$. The topological facts we have developed for the simplified point of view (from the reference frame of one particle, and also taking a finite sphere around the origin) hold good when we abandon that point of view. In other words, there exist closed paths in the six-dimensional space of points (X_1, X_2) minus the diagonal points $<X,X>$ that cannot be shrunk to a point, but every double closed path can be "untwisted."

Now this topological fact has implications for our quantum mechanical formalism.

In this book, the quantum mechanical function $\Psi(x)$ (x is a point in the configuration space) has always been regarded as *single-valued*, though there is no physical difference between the function $\Psi(x)$ and the function $e^{i\delta}\Psi(x)$ for any "phase" δ. Why? Because, as I point out in Appendix A, if the psi-function were multiple-valued, we could find a loop in the configuration space, such that traversing the loop would cause the function to switch values, i.e., add a phase factor. Then we could shrink this loop to a point, making the function switch values in an infinitesimal loop, violating continuity.

But in the projective (configuration) space now under consideration, not every loop can be shrunk to a point—only a double loop can. So it is *consistent* (though, indeed, *not* mandatory) for the function $\Psi_{id}(X_1, X_2)$ to have more than one value.[11] Namely, its value can be multiplied by a complex number c with modulus 1. On the other hand, if it goes around the same loop again, it must come back to its original value. In other words, we have the equation

(2) $$c^2 = 1,$$

which implies that

(3) $$c = \pm 1.$$

These two solutions now explain our two kinds of particles, and in particular the behavior of electrons. Allowing two electrons to switch places (as the result of a continuous motion) is, from the point of

[11] Reread carefully the discussion in Chapter 2 on this point.

view of one of them, like allowing the other electron to move diametrically opposite to it, 180°. But in our projective space, this is an entire loop, not half a loop—and it is consistent for the Ψ_{id} to change sign as a result of the loop.

To clarify this, let us return to our experiment with the two guns. Because of the conditions of the experiment, where the particles always end up 180° apart, switching the two (destinations of the) particles is like rotating our "laboratory" by 180° in Euclidean space. But in projective space, where we identify $<\theta_1,\theta_2>$ with $<\theta_2,\theta_1>$, rotating the laboratory 180° is like coming back to the starting point, i.e., one loop. Consider the series

(4) $\Psi_{id}(90°,270°),\Psi_{id}(135°,315°),\Psi_{id}(180°,0°),\Psi_{id}(215°,45°),$
     $\Psi_{id}(269°,89°),\Psi_{id}(270°,90°)\ [=\Psi_{id}(90°,270°)],$

which are some of the values obtained by "rotating the laboratory 180°," in other words, checking the amplitude of the state in which one particle is at 90° and the other is at 270°, and moving around until we are checking the state where one particle is at 270° and the other is at 90°. Naturally the first and last state must be the same. Yet the function $\Psi(\phi_1,\phi_2)$ need not come back to the same value. It is logically possible, given the topological considerations mentioned above, for $\Psi(269°,89°)$ to be *almost* $-\Psi(90°,270°)$, and for the psi-function to change sign as it continuously makes a loop (not in Euclidean space, but) in projective space.

Now let us apply the Superposition Principle, which states that, when vectors Ψ_A and Ψ_B tell us what the state of a system is in circumstances A and B, respectively—then, when we cannot tell whether the system is, in fact, in circumstance A or B, we *add* the vectors (not the probabilities).

For the case of our experiment, since there is no way to tell whether it is particle 1 at 90° and particle 2 at 270° or the opposite, we should get

(5) $\Psi_{id}(90°,270°) = \Psi_{nonid}(90°,270°) + \Psi_{nonid}(270°,90°).$

But this ignores one critical fact *about the formalism*: if a vector describes a physical system, then the vector multiplied by any complex number of "length" 1 will describe that very system. Thus, equation (5) is too restrictive, and we should write instead

(6) $\Psi_{id}(\mathbf{90°,270°}) = \Psi_{nonid}(90°,270°) + c\Psi_{nonid}(270°,90°), |c|^2 = 1.$

But the only way for $\Psi(\theta_1,\theta_2)$ to change sign (as the position changes continuously from $<\mathbf{90°,270°}>$ to $<\mathbf{270°,90°}>$) is the value

$$c = -1.$$

We have, therefore, arrived at the other possibility: namely,

(7) $\Psi_{id}(\mathbf{90°,270°}) = \Psi_{nonid}(90°,270°) - \Psi_{nonid}(270°,90°) = 0.$

Information contained in the formalism has solved the problem. More generally, of course, we get

(8) $\qquad \Psi_{id}(\mathbf{X_1,X_2}) = \Psi_{nonid}(x_1,x_2) \pm \Psi_{nonid}(x_2,x_1),$

as our two possible applications of the Superposition Principle.

We now, by the way, have a simple answer to the question: what is the value of the psi-function of two identical particles on the diagonal? Namely, zero, or double the value of the psi-function of two *non*-identical particles to be on the diagonal.[12] For in equation (8), the right side obviously goes to zero or to double when $|x_1 - x_2| \to 0$. Hence, if we want $\Psi_{id}(\mathbf{X_1,X_2})$ to be continuous, we require

(9) $\qquad \Psi_{id}(\mathbf{X,X}) = \left\{ \begin{matrix} 0 \\ 2\Psi_{nonid}(x,x) \end{matrix} \right\}.$

That the psi-function of two identical particles should change sign as a result of their switching place in Euclidean space (equivalently, that the psi-function should change sign as a result of one closed loop in the PPS) is not required, only possible. To the contrary, it might not change sign at all. It is reasonable, however, to link this behavior with the nature of the particle.[13] All pairs of electrons change sign; no pairs of pions do.

In sum, then, answering the formal question, "Under what condi-

[12] Recall, though, that in our example, the amplitude for particle 1 and particle 2 to be on the diagonal is zero even for nonidentical particles, because of the nature of the experiment and conservation of momentum.

[13] In fact, all electrons have an intrinsic "spin" or angular momentum that is one-half Planck's constant. (Meaning: if we think of "spin" as a vector, then its z-component is always found to be $\pm\frac{1}{2}$ Planck's constant.) In 1940, Pauli showed that the "antisymmetric" behavior of electrons is explained by their spin.

tion *must* the state vector be determined uniquely?" turns out to have singular physical implications. Of course, we needed the Principle of Superposition, but this, too, is a formal principle.[14] (Why should it make a difference whether we "know" which of the two holes the particle goes through? And so on.)

A skeptical response to the above analysis is: I may be "reading into" the formalism results already known to be true. I shall now present Wilczek's hypothesis, which can certainly not be regarded as ad hoc—it could even be wrong. The fact that Wilczek makes it, though, illuminates the cast of mind that dominates theoretical physics in our century. And, of course, it may well be true.

Let us return, now, to our Punctured Projective Sphere. We discovered that the PPS has topological properties that differ markedly from the Euclidean punctured sphere (ball). In the latter, any closed loop can be shrunk continuously to a point. We say that the Euclidean (punctured) sphere is simply connected. The PPS, though, is not simply connected, because there are closed loops that cannot be so shrunk. On the other hand, any double loop is isomorphic to an ordinary loop in the Euclidean sphere, and can be shrunk to a point.[15]

Let's consider a much easier problem. What does the Punctured Projective Disk look like? In other words, let us consider the case of identical particles confined to a plane. As before, let us consider matters from the reference frame of one of the particles which serves as the origin. Thus we need only two coordinates to locate the other particle. As before, we identify points in the disk with their reflections through the origin: (x,y) with $(-x,-y)$. But, because we are dealing

[14] If any physicist could find intuitive motivation for the Principle of Superposition, it would be Feynman; but, concerning this very principle, Feynman says: "I find it quite amazing that it is possible to predict what will happen by mathematics, which is simply following rules which really have nothing to do with the original thing" (Feynman 1967, 171). This situation could change in the future, of course, if the quantum formalism is embedded in a more general physical theory. Since such a theory has not achieved general currency, my remarks stand, as history and even as contemporary description.

[15] In the language of algebraic topology we say: the Fundamental Group of the Euclidean sphere is trivial (it contains only the zero element), while that of the PPS is that of the integers mod 2, i.e., $Z/2Z$. We regard loops as equivalent, if they can be deformed continuously into one another (homotopy equivalence); the loop equivalent to a point is the zero element of the group. The "multiplication operation" of the group is simply the composition of the loops.

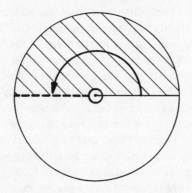

FIGURE 6. The Punctured Projective Disk.
A punctured disk is cut in half, then sewn back as shown, the radii identified; we get the punctured disk back again.

with only two dimensions, and we are three-dimensional beings, it is almost trivial to imagine the PPD. Take the punctured disk, slice off one half of it, and "sew it up" along its diameter in such a way that, on the diameter, diametrically opposed points coincide. Clearly, we get the punctured disk right back again (see Figure 6).[16] Consider an infinite series of loops around the missing center of this disk: the loops go once, twice, thrice, . . . around the center before coming back to their beginning point. Clearly, one cannot continuously deform any of the loops into a point. The implication of this is that a psi-function at a given position can have infinitely many values. Each time the psi-function returns to the "same" position in (configuration) space, it *can* (does not *have* to be, but can) be multiplied by a phase factor (a complex number of modulus 1). This phase factor need not, of course, be -1 (or $180°$), as in the case of ordinary electrons. Instead, it can be *any* rational phase factor. And this has acute implications for our experiment. In our scattering experiment with two electrons, we treated the problem as a two-dimensional one: we

[16] This way of imagining projective space fails in three dimensions, however. The PPS could be thought of as what you get by taking the (punctured) sphere, lopping off the southern hemisphere, and sewing things back up so that opposite points coincide. This operation cannot be pictured in three dimensions.

had detectors at a fixed distance r from the center of the collisions, and we measured only the "longitude" angle ϕ of the electrons, not the "latitude" angle θ. Thus, only the plane of the "table" concerned us, though we well knew that the electrons could fly up in the air or down to the ground. (If one flies up, the other flies down.)

But though the third dimension, altitude, is only a silent spectator in this experiment, its mere existence influences the result profoundly. For if the laboratory—the world—*were* really two-dimensional, the electrons would no longer "live" on the PPS, but rather the PPD. In other words, they would no longer have the property that the state vector describing two electrons changes sign from $+$ to $-$ when those two electrons change positions (i.e., when, in our experiment, the laboratory is rotated 180°). Instead, the state vector could change by any phase at all—the probability that *some* particle should end up going due north could then be any value *between* the cases of fermions (zero) and that of bosons. Identical particles, in a word, could be "anyons," as the current jargon goes.

There is, to be sure, an argument, popular in textbooks, according to which this conclusion is impossible. This is the argument that the only two kinds of identical particles are bosons and fermions. It goes as follows.

Consider the psi-function, $\Psi(x_1, x_2)$, for two identical particles. Make the assumption that there is no way to distinguish between the particles, and that the formalism should reflect that fact. Thus interchanging the particles is nothing more than interchanging their names, and the functions $\Psi(x_1, x_2)$ and $\Psi(x_2, x_1)$ describe exactly the same state. Since multiplying a quantum state function (vector) by a phase factor does not change the described state at all, we are allowed to write, in general,

$$(10) \qquad \Psi(x_2, x_1) = c\Psi(x_1, x_2), \ |c| = 1.$$

Interchanging the variables twice brings us back to our starting point, whence

$$(11) \qquad \Psi(x_1, x_2) = c^2\Psi(x_1, x_2)$$
$$\Downarrow$$
$$c^2 = 1, \ c = \pm 1.$$

That is, there are only two possibilities: the state function must be either symmetric or antisymmetric. Since the argument did not depend upon the dimension of the configuration space, there cannot be "anyons" in any dimension.

There is a subtle fallacy in this argument, however.[17] It is true that if we permute the variables of a function *term* $\Psi(x_1, x_2)$ twice, the *term* will come back to itself. But from this it does not follow that if we switch the *particles* twice, the state must come back to itself. Suppose the particles are situated so that we can switch them simply by rotating our measuring instrument and we do this twice. There is no more reason to think that the twin particles will come back exactly to their original state than there is to think that a single electron, rotated 360°, will come back to itself. (Which, we have seen, does not happen.)

In fact, the same kind of consideration which allowed for the possibility that, when the particles are switched once, the function, though it continues to describe the "same" state, undergoes a phase change, must allow for the possibility that when the particles are switched twice, the function may *not* regain its original phase.

The mistake in the accepted argument, then, is the confusion between the sign and the things symbolized. What is striking however, is how little this confusion matters. For, as it happens, as we have seen, the conclusion of this fallacious reasoning is perfectly correct— in three dimensions. Why need we worry about two dimensions, when electrons are happy to reside in three?

Enter Wilczek. He conjectures that electrons, trapped in a thin layer of material, behave like the two-dimensional particles they are not. They lose their fermionic identities, and masquerade as anyons. This extremely bold hypothesis has both predictive and explanatory value. For example, Wilczek uses it to explain some of the perplexing phenomena of solid state physics, like superconductivity.

Wilczek's thesis is very controversial, because, obviously, no "thin layer of material" is two-dimensional. Electrons, on principle, cannot even be confined to a plane by force—says Heisenberg's Uncertainty

---

[17] The refutation that follows is inspired, of course, by Wilczek 1991. Wilczek's ideas, however, are couched in the language of Feynman path integrals, where I have preferred to avoid this terminology.

Principle. (This is one of the reasons why the Bohr model of the atom, which is a two-dimensional Keplerian system, fails.)

On the other hand, dimensional reduction is not new in physics. Given the right conditions, a three-dimensional system can behave like a zero-dimensional one. Thus, Newton showed that two perfect spheres attracting one another by an inverse square law would behave as though their masses were concentrated in the centers of the spheres. Wilczek in fact has postulated a physical mechanism to explain how electrons might adopt the behavior, "virtually," of particles in a plane. This mechanism is that of the Aharonov–Bohm effect,[18] in which, in the two hole experiment, a solenoid (electromagnetic coil) is interposed between the wall with the holes and the screen behind. Turning on the solenoid changes the interference pattern on the screen, even though outside the solenoid there is no electromagnetic field whatsoever. (The secret is that the electromagnetic *potential* is not zero even though the fields are.)

For our present purposes, it is sufficient to say that turning on a solenoid is mathematically equivalent to changing the topology of the configuration space. The solenoid can destroy, electromagnetically, the fermionic behavior of the electron, so that no matter how many loops one takes around it, the phase of the electron never comes back to itself. Effectively, the problem becomes a two-dimensional problem. Thus, Wilczek, controversially, associates a little "solenoid" to go along with every electron in the kind of solid state problems he is concerned with solving.

Should Wilczek turn out to be correct, the method of research by studying one's formalism, rather than "the world," will gain a new and wonderful vindication.

It is now time to sum up this chapter.

One of the most powerful methods of discovery in physics, particularly in quantum mechanics, is to attempt to extend the formalism to cover new cases. For it turns out that the formalism itself gives hints at its own extension.[19] More than that: the extension of the quantum mechanical formalism is often "forced," in the sense that

---

[18] I discussed this effect in Steiner 1989, 477–9.

[19] I have not studied another example of a formalism, the thermodynamic formalism. I am told that a similar phenomenon exists there.

the extension of the notation for exponent is "forced," when mathematicians defined exponentiation for real and even complex exponents. This ideas can provide retrospective "derivations" for known results, as when we derive the quantization of angular momentum. Thus, although these results could have been discovered by formal considerations, they were not. Once physicists realized the potential of the method, however, they started using formal extension in an attempt to conjecture as yet unknown results.

Now, reasoning by formal extension is certainly Pythagorean. But it is also what I have called "formalist," though mildly so, compared to what I will present in the next chapter. In reasoning by extension, one is adapting a language, a syntax, to new uses. I say "a language," rather than "a mathematical structure," because before the extension takes place, the formalism is simply not defined for the cases in question.

In this chapter, I have presented only one real conjecture based on formal extension (the other examples were reconstructions of known results by formal extension)—namely, the extension of the quantum mechanical formalism to configuration spaces with "deviant" topologies, such that not every loop can be shrunk to a point. There is no reason to believe—except by a formal analogy to other cases of extension—that the extension of the formalism in this way should continue to be "empirically adequate." Physicists who nevertheless engage in such reasoning, I would argue, have abandoned naturalism.

6

Formalist Reasoning: The Mystery of Quantization

Perhaps the most blatant use of formalist reasoning in physics was the successful attempt by physicists to "guess" the laws of quantum systems using a strategy known as "quantization." This strategy begins by assuming that the system obeys the classical laws—a false assumption, of course. Then the classical description is converted (by syntactic transformations) into what is hoped is a true quantum description of the same system. In this chapter, I will show that the discoveries made this way relied on symbolic manipulations that border on the magical. I say "magical" because the object of study of physics became more and more the formalism of physics itself, as though symbols were reality—and the confusion of symbols with reality is what characterizes much of what we call magic. I have no great love for this word, however, and readers whom it offends can ignore it safely.

I will begin with the development of Schroedinger's equation—or equations—for atoms. I will allow myself—in the interests of clarity, if not historical accuracy—to describe his achievement in mathematical terms not necessarily known to Schroedinger himself.

The state of a quantum system, we have seen, is represented by a *single* unit vector in a complex Hilbert space (again, I emphasize, these are not Schroedinger's terms). Schroedinger's equation describes the movement of this vector through time. Since the vector

My thanks to Shelly Goldstein for lengthy discussion and voluminous correspondence on the subject of this section. The title of this chapter is taken from one of his lectures.

is to remain a unit vector, the "movement" must be what mathematicians call a "unitary transformation." This is a transformation of a complex space that preserves the "length" (norm) of every vector; (invertible) unitary transformations replace rotations in complex spaces.[1]

Thus, if $U(t)$ is a group of unitary transformations such that $U(0) = I$, and the system at $t = 0$ is represented by the state vector Ψ_0, then the system at any time t is represented by $U(t)\Psi_0$. Naturally, the identity of $U(t)$ depends on the forces at work in the system under discussion.

We begin by noting that $U(t)$ is a group of unitary transformations such that $U(0) = I$ if and only if there exists a self-adjoint[2] transformation H for which the following equation holds:

$$U(t) = e^{-\frac{i}{\hbar}Ht}.^3$$

Therefore, our state vector can be represented at any time by the equation

$$\Psi(t) = e^{-\frac{i}{\hbar}Ht}\,\Psi(0).$$

In short, to find the state of our system at any moment, it is enough to know the value of a certain self-adjoint transformation H. This transformation is called the "Hamiltonian" of the system.

---

[1] The mathematical definition is: U is a *unitary* transformation if $UU^* = U^*U = I$. (U^* is the adjoint of U: if U is represented by a matrix, then the adjoint is represented by the matrix obtained by simultaneously interchanging rows for columns and replacing each complex number $x + iy$ by its complex conjugate $x - iy$. It is easy to show that unitary transformations preserve the "length" of vectors. I is the identity transformation.)

[2] A *self-adjoint* transformation H on a linear space with scalar product satisfies $\langle x \mid Hy \rangle = \langle Hx \mid y \rangle$. In the finite-dimensional case, this amounts to $H^* = H$, so that if H is represented by a matrix (c_{ij}), self-adjointness is equivalent to the equation $c_{ij} = c^*_{ji}$. For basic facts about transformations, especially the "double role" of self-adjoint transformations as observable and infinitesimal symmetries, cf. Guillemin and Sternberg 1990b, 1–15. All of the mathematical facts cited in the following discussion of the relationship between quantum mechanics and classical mechanics are proved there.

[3] Where transformations are used in exponential functions, the meaning is that the function should be expanded as an infinite power series in the transformation. Represented as matrices, it is not hard to define what we mean by the convergence of such series.

It is obvious from the previous equation that our state vector obeys the differential equation

$$\frac{d\Psi}{dt} = -\frac{i}{\hbar} H\Psi(t),$$

or, multiplying both sides by $i\hbar$,

$$i\hbar \frac{d}{dt}\Psi(t) = H\Psi(t).$$

By simply erasing $\Psi(t)$, we can write this more simply as an operator equation:

$$i\hbar \frac{d}{dt} = H.$$

Our question now becomes: what is H?

Schroedinger's formalist answer—for a nonrelativistic system—was the one accepted by the physics community by the end of 1926. To derive a quantum equation for a system (e.g., an atom), one pretends that the system obeys classical mechanics, writes down the classical energy equation true of such a system, and then "quantizes" the equation. This is done by substituting the quantum operators for the corresponding variables in the equation, arriving at the quantum Hamiltonian H.

Now, the classical energy equation for a system of particles has the form

Energy = Kinetic Energy + Potential Energy,

where the classical kinetic energy is given by

$$K.E. = \frac{1}{2m}(p_x^2 + p_y^2 + p_z^2)$$

for each particle of mass m and momentum $\mathbf{p}$. The potential energy is a function (e.g., a polynomial) of the position(s) of the particle(s) of the system and perhaps the time (where energy is not conserved).[4]

---

[4] Of course, the energy of the entire universe, physicists believe, *is* conserved; but it is often convenient to consider a part of the universe as though it were a separate system under the influence of an outside potential which may vary with time.

In this equation, to derive H we make the following substitutions for each particle:

$$E \rightarrow i\hbar \frac{\partial}{\partial t}$$

$$p_x \rightarrow -i\hbar \frac{\partial}{\partial x}$$

$$p_y \rightarrow -i\hbar \frac{\partial}{\partial y}$$

$$p_z \rightarrow -i\hbar \frac{\partial}{\partial z}$$

$x \rightarrow$ multiplication by x
$y \rightarrow$ multiplication by y
$z \rightarrow$ multiplication by z.

Also, wherever a variable (like momentum) appears classically raised to a power, we iterate the corresponding operation the appropriate number of times. Thus, we get, for example,

$$p_x^2 \rightarrow -\hbar^2 \frac{\partial^2}{\partial x^2},$$

the minus sign the result of multiplying twice by i. A trivial, but useful, remark is: if $V(x)$ is a function of x, and if Q_x is the operator that multiplies a function by x, then $V(Q_x)$ is the operator of multiplication by $V(x)$.

For the case of one particle in an external field, if we write $\Psi(x,y,z,t)$ for the spatial coordinates of our state vector Ψ at any time t, we get the partial differential equation

$$i\hbar \frac{\partial \Psi(x,y,z,t)}{\partial t} = \left[-\frac{\hbar^2}{2m} \left(\frac{\partial^2}{\partial x^2} + \frac{\partial^2}{\partial y^2} + \frac{\partial^2}{\partial z^2} \right) + V(x,y,z) \right] \Psi(x,y,z,t)$$

The result is known as the "Schroedinger equation" for that system.

Ignoring relativistic effects, and to a reasonable approximation, we can regard the hydrogen atom, classically, as a "planet" rotating around an immovable "sun"—the much more massive proton. The "gravity" holding the electron in orbit is the Coulomb attraction between the equal but oppositely charged proton and electron. Thus the electron feels a potential proportional to

$$-\frac{e^2}{r}$$

where $r = \sqrt{x^2 + r^2 + z^2}$ is the distance of the electron from the nucleus (the proton). Thus Schroedinger's equation for the hydrogen atom is obtained from the above by the substitution

$$V(x,y,z) = k\frac{e^2}{r},$$

k an empirically determined proportionality constant.

Thus we have determined the coordinates of the state vector at any time by pretending that the hydrogen atom obeys Kepler's laws and then "quantizing" this absurd assumption away. In quantum mechanics, of course, the very idea that the electron has both a momentum and a position is meaningless—thus the classical Hamiltonian, though mathematically a well-defined function, is a physically meaningless expression.

Nevertheless, the Schroedinger equation for the hydrogen atom worked. The next step was to generalize the method to the helium atom and thence to the heavy atoms. Consider the case of helium—which has 2 protons and 2 electrons. Classically (and nonrelativistically) the energy of a helium atom whose nucleus is stationary is equal to the sum of:

A. The kinetic energies of the electrons;
B. The (negative) potential energy of the electrons from being attracted to the nucleus;
C. The (positive) potential energy of the electrons from repelling each other.

In this case, we must take into account the coordinates of both electron 1 and electron 2, and the Schroedinger equation becomes

$$ih\frac{\partial\Psi(x_1,y_1,z_1,x_2,y_2,z_2,t)}{\partial t} =$$

$$\left[-\frac{\hbar^2}{2m}\left(\frac{\partial^2}{\partial x_1}+\frac{\partial^2}{\partial y_1}+\frac{\partial^2}{\partial z_1}\right)-\frac{\hbar^2}{2m}\left(\frac{\partial^2}{\partial x_2}+\frac{\partial^2}{\partial y_2}+\frac{\partial^2}{\partial z_2}\right)\right.$$

$$\left.-k\frac{e^2}{r_1}-k\frac{e^2}{r_2}+k\frac{e^2}{r_{12}}\right]\Psi(x_1,y_1,z_1,x_2,y_2,z_2,t),$$

where m is the mass of each electron; r_1 and r_2 the distance between electron 1 and electron 2 from the nucleus, respectively; and r_{12} is the distance between the two electrons.

Was there any reason, before the fact, to expect that quantization, which had worked for the hydrogen atom,[5] would work for helium? The expectation rested on three supports.

(a) Performing the indicated substitutions assures that the Schroedinger equation so derived will, in the "classical limit," go over into the classical equations of physics.[6]

What this means is only that, whenever Planck's constant can be regarded as negligible (as in the case of most macroscopic phenomena), then *to a high degree of probability* the observed behavior of a system will be governed *approximately* by the classical equations. Contrast this with the case of special relativity where, as the speed of light is thought of as approaching infinity, the classical equations become, with certainty, better and better approximations to the truth. Thus, classical mechanics is not straightforwardly a "limiting case" of quantum mechanics.

But even if we accept the standard claim that (in some sense) Schroedinger's equation has Newtonian mechanics as a limiting case, this is no argument that Schroedinger's equation is true where Newtonian mechanics is manifestly false.

(b) For the free particle (where $V = 0$), Schroedinger's equation reduces to the de Broglie theory of matter-waves, for which there are physical arguments, buttressed by Einstein's theory of relativity.[7] Thus Schroedinger's equation is just a generalization of the de Broglie theory for the case of arbitrary potentials.

---

[5] It had also worked for the harmonic oscillator.

[6] For the mathematical details of this, see Landau and Lifshitz 1965, § 17.

[7] Cf. Weyl 1950, 48–54 for a concise account of this theory. Schroedinger's equation is, of course, non-relativistic; but the notion that, if the time derivative of a wave (function) is (proportional to) its energy, then its space derivative is (proportional to) its momentum, is both logically and historically motivated by Einstein's special theory of relativity.

This is just the kind of reasoning this book has been written to expose. *Formally*, this reasoning is correct. But *physically* there are reasons to spurn it. For simplicity, consider a single particle moving along the x-axis only. Then we can drop subscripts and write simply,

$$\text{Energy} = \frac{p^2}{2m} + V(x).$$

Substituting operators for these variables, as above, we get an equation that has the following form:

$$\hat{H} = \frac{\hat{p}^2}{2m} + V(\hat{Q}).$$

What this equation means is that the left side of this equation and its right side give the same result when applied to the coordinate function $\Psi(x)$. The equation would make sense—indeed, it would express the de Broglie theory—if it asserted, or implied, that a particle with a definite total energy state can be represented as having a definite kinetic energy and a definite potential energy so that the energy is precisely

$$\frac{p^2}{2m} + V(x).$$

But the potential energy of a particle is a function of its position; the kinetic energy, of its momentum—and, as we know, no particle can have, simultaneously, a definite position and a definite momentum. What this means is that a particle with a definite kinetic energy may have an indeterminate potential energy and vice versa. Or, a particle with a definite total energy may have both its kinetic and its potential energies indefinite, according to the Heisenberg Uncertainty Principle. (In the de Broglie wave theory, by contrast, position does not arise; there is no reason why a free particle cannot have, simultaneously, a definite energy and a definite momentum.) What this means is that there is a colossal difference between Schroedinger's equation without potential energy and the equation with potential energy. The physical motivation for the equation breaks down—Schroedinger's equation is, then, only a formal generalization of the de Broglie theory. As Weyl puts it, "We assume with *Schroedinger* that in spite of the fact that all our variables do not commute we still apply our rules for formation of the wave equation."[8] That is to say, Schroedinger's equation is derived formally from the classical equation by formal substitution of operators even when the original motivation for the substitution no longer applies.

Ultimately, then, the true argument for Schroedinger's equation is:

(c) The success of quantization in the case of hydrogen argues for its probable success in the case of helium (and the other atoms).

But here the question is: what is "it"? That is, are we sure that quantization in the case of helium is the "same thing" as in the case of hydrogen? Yet this question is nothing but the question of whether the following hypothesis is a projectible one:

(H) The Schroedinger equation arrived at by quantizing a classical atom gives a good (nonrelativistic) description of that atom.

For one cannot separate questions of "identity" from questions of projection.[9] Earlier, I pointed out that, for the naturalist, anthropocentric hypotheses are "unprojectible." For this very reason, one must conclude, that, for the naturalist, anthropocentric *analogies* are invalid. That is, we cannot (if naturalists) argue that we are simply "doing the same thing" when the criterion for "same" is an anthropocentric one.

Suppose, then, that the "naturalist" physicist argues—with Weyl, above—that we are merely following "the same procedure" as before. Hence the expectation, based upon past successes, that our equation for helium is correct, is rational. But Weyl's argument is not available

[8] Weyl 1950, 56. Compare Landau and Lifshitz 1965, 51: "The form of the Hamiltonian for a system of particles which interact with one another cannot be derived from the general principles of quantum mechanics alone."

[9] As I have emphasized throughout this book, this point is made with great force by none other than Ludwig Wittgenstein in his *Philosophical Investigations* (Wittgenstein 1968). Speaking in the context of rule-following, he argues that whether one is, or is not, "doing the same thing as before" depends simply upon which rule one is following.

to the naturalist, since all that is "the same" here is our manipulation of a meaningless expression. (That is, quantum mechanics assigns no meaning to the classical energy equation

$$E = \frac{p^2}{2m} + V(x);$$

though, formally, the equations of quantum mechanics "look like" the classical equation the more Planck's constant becomes negligible.) The analogy here is anthropocentric, because notation-dependent.

For heavy atoms, of course, the problem is intensified. It is generally asserted that classical chemistry is, to a good degree of approximation, "nothing but" Schroedinger's equation as applied to the electrons.[10] That is, chemistry is understood to be reducible, more or less, to the theory of electrons moving in the attractive Coulomb potential of a positive nucleus, and repelling one another with the same Coulomb force. But since nobody has demonstrated that this dogma is actually true,[11] the ground for this widespread belief can only be those successes of Schroedinger's equation given in the textbooks. The truth is, that even for helium, the Schroedinger equation is not solvable analytically—and for the heavier atoms, one can rely only on various techniques of approximation. Thus, if Schroedinger's equation cannot be grounded in its own physical picture,[12] its confirmation (in the case of heavier atoms) rests on its analogy to the equation in the case of the lighter atoms—i.e., upon quantization, and thus upon an anthropocentric analogy. In other words, so long as Schroedinger's equation is allowed to be the basis of (nonrelativistic) quantum mechanics, it is unavailable to the naturalist.

[10] I am ignoring here electronic spin, Pauli's principle, and, generally, relativistic effects.

[11] Cf. Primas 1983. More generally, on the relation between fundamental equations and actual phenomena, cf. Cartwright 1983.

[12] I insert this proviso because there are those—Bohm, Bell, and, recently, Shelly Goldstein—who believe that Schroedinger's equation can be demonstrated from a deterministic physical theory without the aid of quantization. If this is true, then we have an ex post facto explanation of why quantization succeeded. The historical mystery remains, however, since those who developed quantum mechanics using quantization were not aware of, or rejected outright, "Bohmian mechanics," they were led primarily by what amounts to superstition.

Scientists who trust it, therefore, are implicitly going beyond natu-ralism—to any of a number of possible metaphysical alternatives (which one, will depend on the individual).

Radical theses are often misunderstood; so I shall restate mine in other terms. In quantum mechanics, according to its orthodox inter-preters, we dispense with pictures and models. We cannot even speak of the "path of a particle" in quantum mechanics, much less of its "orbit" around a nucleus.[13] Some philosophers even regard this as a good thing, a philosophical advance. All there is to quantum mechanics, then, is the formalism itself. And the formalism has no descriptive role: it keeps track of the probabilities that various "observables" will be measured (classically) to have different possible "values."

Now critics like Bell or Bohm protest here that this point of view robs quantum mechanics of its connection to the rest of physics—indeed, robs it of its very right to be called a physical theory. My own view is more radical than theirs (though compatible with it). We have no basis for saying, for example, that the uranium atom is *relevantly* like the helium atom (our intuition to the contrary being based on a continuing, though tacit and officially denied, commitment to classi-cal physics). Thus the success of the *formalism* of quantum mechan-ics in predicting the properties of helium should have no bearing on its probable success with uranium.

Indeed, if we take seriously the orthodox view that quantum mechanics is nothing but the quantum mechanical *formalism*, the only connection we have between, say, the Schroedinger equation for helium and that for uranium, is itself a formal connection. This con-nection is that both Schroedinger equations are derivable by "quan-tization" from the appropriate classical equations. But to say that the connection is "formal" is just another way of saying that the con-nection is mediated by nothing more than notation. And a connec-tion mediated by notation, I have been arguing, is anthropocentric.

I am not a physicist, and my criticism here is not of quantum mechanics—I leave that to Bohm and Bell. I have no quarrel with a physicist who is happy with the status quo, and works with quantum

[13] Landau and Lifshitz 1965 take this line consistently.

mechanics as a mere formalism. My only claim is that such a "happy" physicist has no right to be a naturalist.

It is often said that the era of quantization is over. That is, one no longer guesses the laws of motion of a quantum system by transforming a classical description to a quantum one. If my analysis is correct, however, the continuing validity of Schroedinger's equation for new cases rests on quantization. Thus the era of quantization will not be over until Schroedinger's equation is derived in a standard way from the first principles of an underlying physical theory, which does not depend upon its formal relationship to classical theories for its formulation.

* * *

Another route to "quantization" was that of Heisenberg, Born, and Jordan[14]—that of "matrix mechanics." To be sure, I have already discussed Heisenberg's ideas as examples of *Pythagorean* analogies. But a closer analysis reveals that many of Heisenberg's analogies were in fact *formalist*.[15] The matrix physicists, by and large, were ignorant of the mathematical basis of their analogies.

For the mathematician, a matrix, as notation, represents a linear transformation of a vector space; *which* one is undetermined, until we fix a *basis* of the vector space. To put it another way, infinitely many equivalent matrices represent the same linear operator. A linear operator, in turn, can represent a symmetry transformation; for example, a five-dimensional, or an infinite-dimensional, matrix might represent[16] a three-dimensional rotation or translation. (This idea is the basis for an entire field of mathematics, group representations.) In short, a matrix is a mathematical notation which represents mathematical objects. It is not the focus of interest by itself.

For Heisenberg and Pauli, however, a matrix was an array of quantized variables, not a transformation or a symmetry proxy. That is, it was a "convenient" bookkeeping device. Substituting matrices for variables in an equation, and interpreting the algebraic and ana-

---

[14] From now on, I will just mention Heisenberg.

[15] Recall that a formalist analogy is also Pythagorean; the converse is not always true.

[16] By "representation" here is meant a homomorphism from the group to the set of linear operations on a vector space.

lytic operations of the equation as matrix operations, yielded, therefore, an equation with the same outward form as the classical original. The analogy was then to the form, the syntax, of the equation, not to the mathematical content. It was thus a formalist analogy.

The formalist nature of scientific discovery is often hidden, because scientists attempt to rationalize it *post facto*—and, as in the case of matrices and group representations, they often discover that the mathematicians have done the work already. My purpose here is to tear away the later rationalizations or "reconstructions," to understand what the quantization program really was.

The program began straightforwardly enough. Heisenberg interpreted the coordinates of both position (x_1,x_2,x_3) and momentum (p_1,p_2,p_3) as *Hermitian*, or *self-adjoint*, matrices X_1,X_2,X_3,P_1,P_2,P_3. These are matrices of complex numbers that return to themselves when rows are transposed to columns *and* each matrix entry $a + bi$ is "conjugated" to $a - bi$. An example of a two-dimensional Hermitian matrix is

$$\begin{pmatrix} 2 & 1+i \\ 1-i & 3 \end{pmatrix},$$

One reason for requiring such matrices is that, as is obvious, the diagonal of a Hermitian matrix consists solely of real (i.e., not imaginary) numbers and thus can express observable physical quantities.[17]

There is, however, a striking difference between matrices and the classical magnitudes they replace: matrix multiplication is not commutative in general; FG may not equal GF. In modern phrasing, the "commutator" $[F,G] = FG - GF$ of F and G is not always zero. Heisenberg laid down the famous relations between the position and momentum matrices:

$$[X_i,P_j] = X_iP_j - P_jX_i = \begin{cases} -2\pi i\hbar, i=j \\ 0, i \neq j \end{cases}.$$

So now we have two constraints on the position and momentum matrices: that they should be Hermitian, and that they should satisfy

---

[17] The other reasons Heisenberg and Pauli gave for Hermitian matrices (physical in character) will not concern us here; they were unaware of the deep mathematical meaning of Hermitian matrices which I will touch upon later.

these commutation relations. Significantly, only *infinite* matrices can satisfy these two conditions.

Fundamental in classical mechanics is the law of conservation of energy,

energy = kinetic energy + potential energy,

expressed mathematically by

$$E = \frac{1}{2m} \| \mathbf{p} \|^2 + V(x_1, x_2, x_3).$$

The right-hand side of the equation (let us abbreviate it $H(\mathbf{p},\mathbf{x})$) is called the "Hamiltonian." From the Hamiltonian, we can derive the equations of motion of the classical system.

Heisenberg's idea was to replace this equation by the matrix equation of the same form:

$$E = \frac{1}{2m} \| \mathbf{p} \|^2 + V(X_1, X_2, X_3),$$

where matrices (represented by capital letters) replace position and momentum coordinates, and the energy E is replaced by a diagonal matrix (i.e., zeros in all places except the diagonal), expressing the different quantized values of energy.[18] Perhaps a better way to describe the idea is that the classical equation would be given a nonstandard interpretation: a matrix interpretation, rather than a numerical interpretation. Note that the classical Hamiltonian does not involve the product of position by momentum, so that commutation problems do not arise when we interpret the variables as matrices.

For example, in the one-dimensional harmonic oscillator, $V(x)$ is the function kx^2. No commutation problems arose for Heisenberg, because of the separation of the variables for position and momentum. All one needed to do was interpret E as an infinite diagonal matrix (giving the quantized values of the energy), and interpret $p_1, p_2, p_3, x_1, x_2, x_3$ as infinite matrices, satisfying the commutation relations

$$[X_i, P_j] = X_i P_j - P_j X_i = \begin{Bmatrix} -2\pi i\hbar, i=j \\ 0, i \neq j \end{Bmatrix}.$$

[18] Energy actually is not always quantized in quantum mechanics; but it is in the "bound states"—such as those of atoms—that Heisenberg was interested in.

This was a great success. Solving the equation gave the correct energy levels for the oscillator.

Trouble arose, however, for the simplest "real" problem—the hydrogen atom. The hydrogen atom was understood as a minuscule solar system, obeying Kepler's laws, with the electron its "planet" revolving around the positive nucleus as "sun" according to Coulomb's law, which is an inverse square law like gravity. Now, what was the matrix analogue to planetary motion? The energy equation was of no use here, because the energy of a planet is given by the equation

$$E = \frac{1}{2m} \| \mathbf{p} \|^2 + \frac{km}{r},$$

where r is the distance of the planet from the sun. If we represent the distance r by a matrix R, we still have the problem of representing $1/R$. We cannot assume that R has an inverse, and therefore even if $1/r$ has a quantum analogue, its relation to the quantum analogue of r remains a mystery.

In 1925, however, Wolfgang Pauli vindicated Heisenberg's faith that matrix mechanics had a description of the hydrogen atom by producing one.[19] Pauli's description rederived known results and predicted surprising new ones.

Pauli's method of quantizing the laws of Kepler avoided the energy equation, concentrating instead on other conserved quantities of planetary motion. As pointed out by Pauli's own department chairman, Lenz, in 1924, classical planetary motion has two conserved quantities, besides the energy, whose conservation is both necessary and sufficient for Kepler's three laws. Furthermore, by knowing these quantities, one can calculate the constant total energy of the planet in orbit.

The first of these quantities is the angular momentum of the planet; a vector, perpendicular to the plane of the orbit, whose algebraic definition is $\mathbf{x} \times \mathbf{p}$, that is,

---

[19] Pauli 1926. The result was obtained a year earlier. An extremely lucid reconstruction of the argument using group theory and Lie algebras is contained in the first six chapters of Guillemin and Sternberg 1990b. I have drawn upon this book freely in my own treatment. All of the mathematical claims I make in what follows are proved there.

$$l_1 = x_2 p_3 - x_3 p_2$$
$$l_2 = x_1 p_3 - x_3 p_1$$
$$l_3 = x_1 p_2 - x_2 p_1,$$

where the xs are position coordinates and the ps are momentum coordinates.

The conservation of angular momentum implies that the orbit of a planet remains in a plane. It also implies the area law of Kepler— the radius vector of the planet sweeps out equal areas in equal times. The conservation of angular momentum arises from a rotational symmetry: the gravitational pull is the same in all directions. To quantize angular momentum, we need only follow Heisenberg's recipe, because the matrices X_i and P_j in the definition all commute.

Lenz, however, discovered a more subtle conserved quantity, one which is needed for the other two laws of Kepler, or, equivalently, for the inverse square law. This is the vector that bears his name; it lies along the axis of the ellipse and its length is equal to the eccentricity of the orbit. The conservation of the Lenz vector means that the orbit does not precess (swing around) or change its shape.[20] The definition of this vector is

$$\mathbf{f} = \frac{1}{mk} [\mathbf{p} \times \mathbf{l}] + \frac{\mathbf{x}}{r},$$

where $\mathbf{l}$ is the angular momentum; $\mathbf{p}$, the linear momentum; $\mathbf{x}$, the position of the "planet"; r, the distance from the "sun"; m is the mass of the "planet"; k, a constant of proportionality. It is not hard to see that the energy can be calculated using the Lenz vector, and in fact,

$$1 - \| \mathbf{f} \|^2 = - \frac{2E}{mk^2} \| \mathbf{l} \|^2.$$

Pauli's idea was to find the quantum analogue of the Lenz vector. An analogue to the classical functional dependence of the energy and the Lenz vector would allow him to solve for the quantized energy of the hydrogen atom. Indeed, Pauli was able to calculate that the nth

---

[20] Interestingly, Newton not only already knew that the nonprecession of the orbit of a planet is necessary for the inverse square law, he used this nonprecession as a sensitive test for the inverse square law. This same test was used later by Einstein to refute the inverse square law, as the orbit of Mercury does precess. Cf. Harper 1990.

energy level of hydrogen corresponded to n^2 different states. This is called "energy degeneracy."

There was one problem, however. The quantum Lenz vector could not be defined merely by interpreting position and (linear) momentum as matrices in the formula

$$\mathbf{f} = \frac{1}{mk} [\mathbf{p} \times \mathbf{l}] + \frac{\mathbf{x}}{r}.$$

The components of the matrix expression $\mathbf{P} \times \mathbf{L}$ contain noncommuting products of the form X_iP_i. For this reason, the expression is not Hermitian. Pauli, whether he realized it or not, was faced with the problem of finding a suitable version of $\mathbf{p} \times \mathbf{l}$ which could be quantized by direct substitution of matrices. Pauli suggested the following symmetric expression:

$$\tfrac{1}{2}(\mathbf{p} \times \mathbf{l} - \mathbf{l} \times \mathbf{p}).$$

This is the same as $\mathbf{p} \times \mathbf{l}$ classically and is well defined.

Unfortunately, there are infinitely many different "symmetrized" matrix expressions that are classically equivalent to the Lenz vector. The only argument that Pauli gives for his choice is this. The commutator of position and angular momentum, can be easily calculated as

$$[X_i,L_j] = X_iL_j - L_jX_i = \begin{Bmatrix} 0, & i=j \\ i\hbar, & i\neq j \end{Bmatrix}.$$

If we use Pauli's quantized Lenz vector, we get the following formally analogous commutation relations:

$$[F_i,L_j] = F_iL_j - L_jF_i = \begin{Bmatrix} 0, & i=j \\ i\hbar, & i\neq j \end{Bmatrix}.$$

So the only argument Pauli gives for his choice of the Lenz vector in matrix mechanics is formalist to the core. This shows that quantization is not (even) an algorithm, but involves faith in the Heisenberg (et al.) prophesy each time we quantize.

The transformation of the Lenz vector by quantization has profound results. For an arresting example, consider again the classical equation

$$1 - \| \mathbf{f} \|^2 = \frac{E}{km} \| \mathbf{l} \|^2.$$

In matrix mechanics, this equation is shifted as follows:

$$1 - \| \mathbf{F} \|^2 = \frac{E}{km} \| \mathbf{L} \|^2,$$

which of course changes the classical energy values, aside from quantizing them.

(This expression, to be sure, approaches the classical expression in the limit, as $\hbar \to 0$. But not every quantum analogue does that. For example, where the (absolute value of) the classical angular momentum is s, the quantum analogue must be $\sqrt{s(s+1)}$, which does not approach s in the limit. To be sure, there are other definitions of "limit"—and it is true that

$$\lim_{\hbar \to \infty} \sqrt{\frac{s(s + 1)}{s}} = 1,$$

which is a weaker form of limit.)

Historically, the energy levels of hydrogen (the "Balmer series") had already been derived by Bohr. Thus a cynic might argue that Pauli reasoned backwards, choosing the form of the Lenz vector that would give the Balmer series.

The answer to the cynic is, first, that *Heisenberg's* formalist prediction is still vindicated—he predicted that it would be possible to find a quantum analogue of any classical problem, and this turned out to be true for the hydrogen atom.

Second, Pauli's analogue to the Lenz vector also led to a new and successful prediction: that the first energy level of hydrogen would have to correspond to angular momentum zero. This result seemed physically absurd because if the electron were not orbiting the nucleus, it could not produce magnetism, and the so-called alkali metals, physically analogous to the hydrogen atom, would certainly be counter-examples. Pauli stuck by his result, then, even though it seemed to defy the facts. And he was right: the magnetism of the alkali metals does not stem from orbital angular momentum, but from a special intrinsic magnetism of the "spinning" electron itself.[21]

---

[21] It is interesting that Pauli knew about the hypothesis of spin, because it had already been offered by Uhlenbeck and Goudsmit, but was not particularly enthusi-

My claim, then, is: the lack of an algorithm to "quantize" classical systems makes the analogy between classical and quantum mechanics distinctly formalist.

An objection to this is that Pauli's choice for Lenz vector analogue makes a great deal of *mathematical* sense. Pauli's definition ensures that the quantum hydrogen atom has the same symmetries as the classical planetary system.

The two conserved quantities of planetary motion, angular momentum and the Lenz vector, arise from symmetries of the planetary system. Angular momentum conservation arises from rotational symmetry, in which the energy of the system (potential and kinetic) does not change under rotations of space where the sun is at the center. The Lenz vector conservation also arises from a symmetry, but not one involving a transformation of the three dimensions of space. The classical planet "lives" in a portion of a six-dimensional space—called "phase space"—where the coordinates are those of space and those of linear momentum. Lenz vector conservation arises from a transformation of this phase space, not physical space.

In quantum mechanics, the wave function of the planet (i.e., the electron of the hydrogen atom) "lives" in an infinite dimensional vector space, known as a Hilbert space. Abstractly, the same symmetries acting on the classical phase space may act on Hilbert space. In the case of quantum mechanics, the members of a symmetry group are represented by infinite matrices or linear operators; in classical mechanics, they are represented by functions on phase space. In both cases, the functions representing the rotations follow the abstract "multiplication table" of the rotation group.[22]

We can now rephrase the analogy between classical and quantum mechanics as follows: for each conserved quantity in a classical system, there is a conserved quantity in the analogous quantum system such that both quantities arise from the same symmetry group. This "group-theoretic" analogy provides a mathematical rationale for

---

astic about it. Later, to be sure, the concept of "spin" became central to his "exclusion" principle.

[22] Technically, the functions are homomorphic images of the elements of the rotation group.

Pauli's decision concerning the definition of the Lenz vector in quantum mechanics: through his definition, but not through the alternatives, the Lenz vector conservation law arises in both classical and quantum mechanics from the same symmetries.

To this argument, there are two answers.

First, even if this were so, the analogy would be Pythagorean, even if not formalist—so, in any case, opposed to naturalism. There is no physical reason to suppose that the same abstract groups act in a theory as in the one that replaces it. In fact, there is reason to assume that they do not: one would think that a stronger theory has fewer symmetries. That is, the discovery of new forces in nature tends to break symmetries already given. Besides, the entire concept of "symmetry" as used in modern mathematics is Pythagorean, and has only a partial overlap with the ancient Greek concept of symmetry (balance, proportion).

But, second, and more significant for present purposes, Pauli failed to see the mathematical analogy between classical and matrix mechanics (Dirac later pointed it out). Even had Pauli noticed this analogy, he would not have understood its mathematical significance—*for him*, even this analogy would have been formalist. As I remarked before, Pauli saw a matrix merely as an array of numbers, not as a linear operator.

It is easy to document Pauli's "unmathematical" attitude towards matrices. Pauli writes down the following matrix "equations," most of which we have seen already:

$$\mathbf{L} \times \mathbf{L} = i\hbar\mathbf{L};$$

$$[F_i, L_j] = F_i L_j - L_j F_i = \left\{ \begin{matrix} 0, i=j \\ i\hbar,\ i \neq j \end{matrix} \right\}.$$

$$\mathbf{F} \cdot \mathbf{L} = \sum_{i=1}^{3} F_i L_i = 0;$$

$$\mathbf{l} \cdot \mathbf{f} = 0;$$

$$\mathbf{F} \times \mathbf{F} = -i\hbar \frac{2}{mk^2} E\mathbf{L};$$

$$1 - \| F \|^2 = \frac{E}{km} (\| L \|^2 + \hbar^2).$$

Pauli here remarks that "regrettably"[23] he cannot solve these equations without further assumptions on the kind of solutions he is looking for. He goes on to assume, for example, that not only is the energy matrix E diagonal, but also the matrices $\| \mathbf{L} \|^2$ and L_z are diagonal. Alternatively, he assumes that the matrices L_z, F_z are diagonal along with E.

But this remark shows how primitive, in 1925, was Pauli's mathematical comprehension of matrices. Had he seen a matrix as one of many representations of the same transformation of a vector space, he would have seen that the "assumption" that a matrix is diagonal is just a choice of basis for the vector space. The ability to choose different bases is not "regrettable," but reflects the physical situation as it really is. Each choice of basis is a decision which quantities to measure, because we can't measure them all simultaneously in quantum mechanics.

There is another, more subtle, point. Throughout his article, Pauli writes as though the matrix equations he is generating for the hydrogen atom can be viewed as relations among finite matrices, although officially Heisenberg requires the matrices to be infinite. This deviation from Heisenberg is obvious, particularly when Pauli regards the energy matrix not only as a diagonal matrix (having different entries along the diagonal) but as a scalar matrix, i.e., a diagonal matrix having the same entry all along the diagonal, or, equivalently, a number times the identity matrix. This reflects what is called the "degeneracy" of energy, i.e., the fact that one energy corresponds to a finite number of different states of the atom; in fact, it is this finite number that Pauli wants to calculate, *inter alia*. Pauli even divides by the matrix E, although, not knowing in advance what that matrix is, he cannot know in advance that this is legitimate. He simply assumes, without any real proof, that the equations involving infinite matrices remain physically correct when interpreted in terms of finite matrices, the energy matrix collapsing into a *number E*. Pauli couldn't have known that Schur, one of the great founders of the theory of group representations, had already justified this assumption years before.[24]

[23] Pauli 1926, p. 404 in the English version (Van Der Waerden 1967).
[24] Schur's Lemma states that no matrix, except a scalar matrix, can commute with every $m \times m$ matrix that represents a group, irreducibly, on an m-dimensional vector space.

I conclude, therefore, that Pauli's analogy, given what he knew then, was thoroughly formalist.

Although wave mechanics shortly afterward eclipsed matrix mechanics, I believe that Pauli's approach served as a model for further discovery. After all, Gell-Mann attempted to explain the unexpected degeneracy of hadronic particles by attempting to solve matrix equations of the same kind as Pauli's. A mathematician would say that Pauli and Gell-Mann showed that if one knows the symmetries of a system, one can dispense with equations. This is true only from hindsight: both Pauli and Gell-Mann, as they pursued their avenue of discovery, were ignorant of the appropriate concept of symmetry. The appropriate mathematical structures are called Lie groups and Lie algebras, of which both physicists were uninformed.

* * *

I shall discuss two further episodes of quantization, both due to Dirac: his "quantization" of the electromagnetic field (Dirac 1927), and his famous relativistic equation for the electron, which led to the discovery of the positive electron, or positron (see Schwinger 1958, 82–91).

The quantization of the radiation field was perfectly reasonable, given that quantization itself is reasonable (which it is not, for the naturalist). Dirac proceeded as follows:

(a) He confined the radiation to a "box."
(b) Maxwell's theory then gives the radiation as a superposition of countably many "normal modes."
(c) Each normal mode is equivalent, *formally*, to a harmonic oscillator.
(d) Each harmonic oscillator can be quantized, according to Schroedinger's equation (or Heisenberg's, equivalent, approach).
(e) The quantized field, then, is the superposition of countably many quantum oscillators.

The reasoning is formal: there was no physical or mathematical reason to accept the principle that to quantize a sum, we must sum the quantizations. Those who accept quantization as a natural form of reasoning, though, will see this as a reasonable continuation—

"more of the same." I have already argued that a naturalist, though, cannot help himself to quantization, so that Dirac's procedure here (naturalistically) is just more superstitious behavior; and its success is cause for wonderment at the "coincidence."

* * *

Dirac's formal derivation of the electron equation[25] is a high point of the quantization program, and should amaze even those unmoved by the successes of quantization I have so far retold.[26] I have already discussed this derivation as a general mathematical analogy. Now I would like to return to it as an example of quantization, adding some more details.

The method of formal substitution had led Schroedinger (and others) to a relativistic wave equation, as we have seen.[27] After concluding that the way to derive nonrelativistic mechanics was by substituting differential operators in Hamilton's energy equation

$$E = \frac{p^2}{2m} ,$$

he suggested making "the same" substitutions in the Einstein mass-energy equation

$$E^2 - p^2 = m^2 .[28]$$

Schroedinger, recall,[29] explicitly refers to this as a "purely formal analogy." The result (for a free particle) is:

$$\hbar^2 \left[\frac{\partial^2}{\partial t^2} - \left(\frac{\partial^2}{\partial x^2} + \frac{\partial^2}{\partial y^2} + \frac{\partial^2}{\partial z^2} \right) + m^2 \right] \Psi(x,y,z,t) = 0$$

[25] For Dirac's own reminiscences concerning the relativistic equation for the electron, see Dirac 1977. (Where Dirac discusses the work of other physicists, however, this article is not to be relied upon. Where he discusses his own reasoning, the article is obviously more reliable—but one should always treat with caution accounts written over thirty years after the events they detail.)

[26] I was overoptimistic; Eli Zahar cites (Zahar 1989, 39), albeit briefly, this very example, but remains unmoved by it.

[27] See Schroedinger 1978, 118–20. Pais 1986 lists five other authors who derived the Klein–Gordon equation in the space of half a year in 1926.

[28] This is the correct equation in units where the speed of light is unity.

[29] Schroedinger 1978, 118–19.

or, in operator form (where the hat on top of the variable signifies the corresponding operator),

$$\hat{E}^2 - \hat{p}_x^2 - \hat{p}_y^2 - \hat{p}_z^2 = m^2,$$

known today as the Klein–Gordon equation.[30] There is a standard trick which can transform this equation from a free particle equation into a particle moving under the influence of an electromagnetic field.[31]

This equation had a number of "defects," however. First, of course, even after the above-mentioned "trick" was tried, it did not give the correct energy levels of the hydrogen atom. Today we know that this is because the KGE—as a field equation—describes, not the electrons, but spinless particles, such as pions. Second, because the Einstein equation contains the energy squared (Hamilton's contains only the first power of E), the free particle KGE has "negative energy solutions." Of course, even without quantum mechanics, the Einstein equation has negative energy solutions; the difference is that quantum mechanics allows the transition of a particle from one energy to another—and it was not clear, if the Klein–Gordon equation is correct, why we do not see such a transition in nature.

But for Dirac, however, the KGE had an overriding flaw—it was second-order in the time; that is, it contains the second derivative of time. This meant that knowing the initial state of a system at time zero would not be enough to predict the future development of that system. (One would have to have an extra initial condition—just as in the case of Newtonian mechanics.) And this was a violation, for Dirac, of a fundamental formal feature of quantum mechanics.

True, the Schroedinger equation is first-order in time, but it contains the second derivatives of space. And that spoils the symmetry of spacetime necessary for a relativistic equation. Dirac therefore concluded that one must search for an equation that is first order in all derivatives.

---

[30] I have already disputed the widespread opinion that Schroedinger had written down the Klein–Gordon equation even before he arrived at what we call today Schroedinger's (nonrelativistic) equation.

[31] See Messiah 1962, 884 ff., for details.

This did not mean that the KGE was "false," however. To the contrary, it was arrived at by the "correct" formal substitutions—and did, therefore, express the quantum version of Einstein's energy equation. To the contrary, Dirac argued, the equation he was looking for should imply the Klein–Gordon equation (i.e., every one of its solutions should be also solutions of the KGE) but not the converse.

Dirac concluded that the only way to get a relativistic equation for the electron was to *factor* the Klein–Gordon equation:

$$\hat{E}^2 - \hat{p}_x^2 - \hat{p}_y^2 - \hat{p}_z^2 - m^2 =$$
$$(\hat{E} + \alpha_1\hat{p}_x + \alpha_2\hat{p}_y + \alpha_3\hat{p}_z + \alpha_4 m)(\hat{E} - \alpha_1\hat{p}_x - \alpha_2\hat{p}_y - \alpha_3\hat{p}_z - \alpha_4 m) = 0.$$

The right factor, set to zero, is the Dirac equation. The only problem was that the Klein–Gordon equation does not factor. For in order that the "identity"

$$\hat{E}^2 - \hat{p}_x^2 - \hat{p}_y^2 - \hat{p}_z^2 - m^2 =$$
$$(\hat{E} + \alpha_1\hat{p}_x + \alpha_2\hat{p}_y + \alpha_3\hat{p}_z + \alpha_4 m)(\hat{E} - \alpha_1\hat{p}_x - \alpha_2\hat{p}_y - \alpha_3\hat{p}_z - \alpha_4 m)$$

should hold, the following relations among the alphas must hold:

$$\alpha_1^2 = \alpha_2^2 = \alpha_3^2 = \alpha_4^2 = 1$$
$$\alpha_k\alpha_l = -\alpha_l\alpha_k \ (k \neq l).$$

That is, the alphas anti-commute, and their square is 1. Obviously there are no numbers that satisfy these relations, so the KGE does *not* factor.

Undeterred by mere mathematical impossibility, Dirac argued backwards: since the formal relations expressed by the equations are correct, there must be some consistent interpretation for them. In fact, Dirac was able to find four *matrices*, 4×4 each, which satisfied the formal relations (square being 1 and anti-commuting), and which were also self-adjoint. These could be the coefficients of his equation. (Here is an interesting twist to the quantization story. In prior episodes of quantizations, matrices replaced physical magnitudes. In Dirac's case, matrices replaced coefficients.)

But now the tail wags the dog: in order that the *coefficients* of our operator equation be 4×4 matrices, the *solutions* of the equation have to be 4×1 matrices (i.e., quadruples of numbers),[32] so that each solution of the equation is four solutions. Thus, instead of one psi-function, determining the amplitude of finding the particle at a given point in space, we have a matrix of four psi-functions, called a "spinor." And now the miracle happens: our "spinor" gives us information about completely new phenomena, unknown to classical mechanics—and to nonrelativistic quantum mechanics: spin[33] and anti-matter. For the four psi-functions give the probability, not only of the position of the particle, but also of its spin and charge. That is, the four components of the spinor give the probability of the particle being, respectively, an electron of spin "up," an electron of spin "down," a *positron* of spin "up," and a positron of spin "down."

Actually, this particular interpretation of the Dirac equation took hold slowly. (As we have seen, an equation can be in place long before its "true meaning" is understood.) It was clear that two of the four psi-functions in the solution of the Dirac equation were the "negative energy solutions" that had proved puzzling in the context of the Klein–Gordon equation. The idea that a positive particle in an electromagnetic field could behave like a negative particle with "negative energy"—a positron—had occurred to scientists, but nothing like it had been observed. Hermann Weyl spoke for the consensus when he concluded, falsely, "The solution of this difficulty would seem to lie in the direction of interpreting our four differential equations as including the proton in addition to the electron."[34] Dirac, on the other hand, believed in the equation—and constructed a theory to explain why positrons had never been seen.[35] The positron was discovered by an experimentalist, Carl Anderson, in 1932.

[32] These quadruples are not known as vectors, because they do not return to themselves when space is rotated by 360 degrees. Instead, they are known as four-dimensional spinors.

[33] The "spin" of the electron—the intrinsic magnetic moment of the particle—was discovered in the 1920s even before Schroedinger's equation was written down. Schroedinger's equation does not predict the interaction with the electron spin with the electromagnetic field, because its Hamiltonian is derived directly from that of classical mechanics. Nor does the Klein–Gordon equation which, though relativistic, describes spinless particles (which were discovered twenty years afterwards).

[34] Weyl 1950, 225. Dirac himself was part of this consensus for a while.

[35] Dirac's theory postulated a "sea" of electrons having negative energy—invisible

Dirac's derivation was, thus, doubly formal: he began with the standard substitutions that had launched Schroedinger's equation, and shared Schroedinger's faith that, despite the vast difference between Hamilton's and Einstein's energy equations, a substitution that "worked" in one case should work in the other. This yielded the Klein–Gordon equation. The next step was to factor the equation formally, in order to achieve the double desiderata of a first-order equation and a Lorentz-invariant equation. Having arrived at a formal result, he then—and only then—looked for a consistent interpretation of the equation. The first "half" of his quadruple-spinors, he concluded immediately, showed that electron spin was a truly new phenomenon, the offspring of the marriage of quantum mechanics and relativity.[36] The other "half" of the spinor, corresponding to anti-matter, took longer to interpret, because here a prediction—that of the positron—was involved, and it took courage to predict it.

A sheltered reader might here ask: why did it take courage to predict the positron? Does it take courage to predict the eclipse of the moon? Isn't prediction just a matter of simple deduction from the laws of nature—or what we take to be those laws—plus "initial conditions"?[37]

I have news for such readers. The term "prediction" in physics has, in the last hundred or so years, undergone a meaning shift.[38] Prediction today, particularly in fundamental physics, refers to the assumption that a phenomenon which is mathematically possible

---

because of their very ubiquity. On the other hand, the so-called "Pauli exclusion principle" (which states that two electrons cannot be in the exact same state) would prevent visible electrons from sinking into the negative energy states. On the other hand, it remained a theoretical possibility to raise one of these electrons of energy $-E$ to an energy of $+E$, in which case an electron-positron pair could be produced—the positron corresponding to the "hole" left behind in the "sea." This prediction was spectacularly confirmed. Nevertheless, the Dirac "electron sea" theory has been superseded by the deeper field theoretic approach to quantum electrodynamics.

[36] The irony here is that, from a deeper perspective, "spin" can be derived even from non-relativistic quantum mechanics, as will appear below. Hence Dirac's oft-quoted view that "spin" is a "relativistic" phenomenon was a mistake.

[37] Indeed, this standard view of prediction led Carl Hempel and Ernest Nagel, independently, to the "deductive-nomological" model of *explanation* in science, according to which explanation would be the same as prediction, with the one exception that the explanandum (thing explained) has already happened.

[38] I find it interesting that historians, such as Kuhn, who allege wholesale meaning shifts in physics, failed to notice this one—which is real.

exists in reality—or can be realized physically.[39] Even when the physicist works with a deterministic equation, there is often no question of predicting the future from initial conditions, since the whole problem is whether these initial conditions are physically (as distinct from mathematically) possible. (Recall Einstein's view that faster-than-light travel was mathematically possible but physically absurd.) Many of the "Pythagorean" examples of discovery described in this book involve the "prediction" that, since such-and-such a phenomenon is a solution of an equation, the initial conditions in nature must either exist or be realizable.

Even without an equation, physicists can predict events by the use of symmetries. Here virtually all predictions are of the (nondeductive) "possible implies actual" variety, because symmetry conditions define more what cannot occur rather than what must occur.[40] In particle physics, for example, I cited the example of collision experiments, where the mathematical fact that two structures[41] are isomorphic leads to the physical prediction that the corresponding physical structures can be transformed into one other. There is nothing deductively inevitable about such predictions, which are based on the inchoate assumption that mathematical possibilities are realized.

In short, the concept of "prediction" has itself become thoroughly Pythagoreanized. Pythagorean expectations have become "built in" to the extent that they are called predictions. In the case of Dirac, then, to predict the positron took courage—or faith in mathematics. And the equation which supported this Pythagorean prediction, we have now seen, was "derived" by purely formalist maneuvers.

The knowledgeable reader may object that I have attributed to

---

[39] I have already drawn attention to the resemblance of these ideas to Lovejoy's Principle of Plenitude. The main difference is that Lovejoy's principle involves the belief that possibilities *will* be actualized in the fullness of time, while the physicist's "principle" is more activist in that it holds that the realization of possibilities in physics might involve billions of dollars of machinery. Alternatively, the physicist holds that the possibilities already exist in nature, but it might take billions to discover them.

[40] It is important to remember that equations do not define necessities either. When a physicist predicts the future F on the basis of initial conditions C and laws L, what is necessary is the statement "If C and L, then F." From this, the necessity of F does not follow, unless C by itself is necessary, which is usually not the case.

[41] These structures are, in fact, group representations, so we are talking about the use of symmetries.

Dirac some kind of magical manipulation of symbols, when he was only doing algebra. In advanced algebra, we deal with many structures that differ in their properties from the real numbers. In fact, the algebra that Dirac was working with—the algebra of matrices—had already been discovered, and it is called a "Clifford algebra." In this algebra, but not in the algebra of real numbers, the Klein–Gordon equation factors.

My answer is that there are two concepts of algebra available to mathematics. Algebra may be conceived as the study of structures (groups, fields, etc.) which are the class of models of a set of postulates that serve as definitions of the structures. It is this conception that inspires the objection to my account.

But algebra is also the study of symbolic manipulation.[42] One can present a group, for example, by taking the set of all expressions ("words") that are possible using the identity symbol, the inverse symbol, etc. This set is itself a group—called the "free group." Now by identifying certain of the different "words" by rules, we can create what are called "quotient groups" of the free group. Here we have a structure defined in terms of the syntax of expressions, not in terms of models.[43] The power of algebra lies in the different approaches we make take to the subject. My thesis is that Dirac, in discovering his famous equation, was operating simply on the syntax of his expressions. Physical requirements (first-order, Lorentz-invariance) became purely formal requirements, and lost their physical content, at least for a while. Later, the mathematicians discovered that Dirac had rediscovered the Clifford algebra, and Dirac's discovery was then enriched.

Let me summarize the magical discovery by Dirac, made largely by formalist reasoning. Schroedinger had noted that the Schroedinger equation could be obtained by formal substitution of operators for variables in the classical energy equation

$$E = \frac{\|\mathbf{p}\|^2}{2m} + V.$$

---

[42] Only in algebra, for example, do we have mathematical concepts that have an explicitly syntactic characterization: that of "left inverse" and "right inverse," for example.

[43] Shaughan Lavine pointed this out to me.

Making "the same" substitution in the Einstein energy equation $E^2 - \| \mathbf{p} \|^2 - m^2 = 0$ yielded the so-called Klein–Gordon equation. This (admittedly, relativistic) equation had no known use then, but Dirac factored it *formally* to obtain a (formally) linear equation in time, itself formally analogous to the Schroedinger equation (this last analogy was very important to Dirac, and played a dominant role in the discovery, but physicists no longer regard linearity in time as a necessary feature of quantum mechanics). The formal coefficients of this factored equation could be represented as 4×4 matrices, but this compelled quadrupling the equation, yielding always four solutions. These solutions predicted both electron spin and anti-matter. Dirac thus vindicated the basic predictions of the quantization program— that the key to quantum mechanics was by the route of substituting operators for classical variables in the classical equations. But he took the program a few steps further by his trick of formal factoring of the classical equation, and extending the use of matrices to the coefficients of the classical equations, not just the variables.

There is an amazing sequel to the Dirac story. Dirac describes the thought processes that led to his wave equation in the following way:

> [E]ventually the solution came rather by accident, just by playing with the mathematics. I noticed that if you take the matrices $\sigma_1, \sigma_2, \sigma_3$ describing the three components of spin for a spin of half a quantum . . . then if you form
>
> $$(\sigma_1 p_1 + \sigma_2 p_2 + \sigma_3 p_3)^2$$
>
> you get a very interesting result, just
>
> $$p_1^2 + p_2^2 + p_3^2.$$
>
> You had thus a sort of square root for $p_1^2 + p_2^2 + p_3^2$.
>   Now I needed a corresponding expression for the square root of the sum of four squares. . . . That was a serious difficulty for me for some weeks, until I noticed that there is really no need to keep to two-by-two matrices. . . . One can go to four-by-four matrices, and then one can easily get an expression for the square root of the sum of four squares. (Dirac 1977, 3)

Dirac did not realize, nor did anybody else until two Spanish physicists pointed it out (Galindo and Del Rio 1961), was that the "interesting" result that Dirac had discarded,

$$(1) \quad (\sigma \cdot \mathbf{p})^2 = (\sigma_x p_x + \sigma_y p_y + \sigma_z p_z)^2 = p_x^2 + p_y^2 + p_z^2,$$

could be used, not just as a stepping stone, but for its own sake, in demonstrating that the spin of the electron is already implicit in classical, nonrelativistic mechanics.

For, consider the classical equation for the free particle

$$E = \frac{\|\mathbf{p}\|^2}{2m}.$$

Using (1), Dirac's discovery, one can *factor* the equation in the form

$$(3) \qquad \left(E^{1/2} - \frac{\sigma \cdot \mathbf{p}}{(2m)^{1/2}} \right)\left(E^{1/2} + \frac{\sigma \cdot \mathbf{p}}{(2m)^{1/2}} \right) = 0.$$

This remarkable result shows that the formalism of classical mechanics "knows about" spin.[44] We substitute operators for the magnitudes E and $\mathbf{p}$—simultaneously bringing in the electromagnetic potentials using the "trick" of Schroedinger alluded to on p. 99 n. The result (detailed in Appendix C) is a Schroedinger equation for the "spinning" electron in an electromagnetic field. Since the Pauli matrices $\sigma_x, \sigma_y, \sigma_z$ are two-by-two matrices, though, the psi-function now has two coordinates and has the form of a two-by-one matrix

$$\left(\begin{array}{c} \Psi_{up}(\mathbf{x},t) \\ \Psi_{down}(\mathbf{x},t) \end{array} \right),$$

where the "up" component gives the probability to be at place $\mathbf{x}$ at time t and also have spin $+\frac{1}{2}\hbar$ in the z-direction; correspondingly, for the "down" component.

We can see the success of this quantization, and of the formalist prophecy of Heisenberg more generally, from another angle. The equation discussed in Appendix C is the following:

$$1 - 2m \sum_{r=1}^{3} \left(-i\hbar \frac{\partial}{\partial x_r} - \frac{e}{c} A_r \right)^2 \Psi - \frac{e}{m}\frac{1}{c}\frac{\hbar}{2}\, \sigma \cdot \mathbf{B}\Psi - (E - eV)\Psi = 0.$$

---

[44] Cf. Galindo and Del Rio 1961 for the mathematical details.

This equation contains three terms, each of which describes, in quantum mechanical language, the three kinds of energy an electron has in an electromagnetic field in virtue of its spin. (The Schroedinger equation is always the quantum mechanical version of an energy equation.) The first term corresponds to kinetic energy. The third term refers to the potential energy of the electric field. The middle term,

$$- \frac{e}{m} \frac{1}{c} \frac{\hbar}{2} \sigma \cdot B \Psi,$$

gives the potential energy of the coupling of the spin to the magnetic field (spin makes the electron into a magnet). Remarkably, this gives exactly the right answer (disregarding, once more, some fluctuations requiring quantum electrodynamics to explain).

Historians and physicists know, however, that this correct answer was twice the result that standard quantization predicted, and I would like now to explain why two routes to quantization gave two different results.

The standard way to quantize the energy of a charged particle rotating or spinning in a magnetic field is, naturally, by direct substitution in the classical energy equation. Now, classically, the potential energy of a spinning (or rotating) charged particle with angular momentum coordinates S_x, S_y, S_z in a magnetic field with coordinates B_x, B_y, B_z is

(4) $\gamma \left(\frac{1}{c} \right) S \cdot B = \gamma (S_x B_x + S_y B_y + S_z B_z),$[45]

where γ is a constant known as the "gyromagnetic ratio." It only remains to substitute the appropriate operators for the angular momentum coordinates. The gyromagnetic ratio "should" (if we believe in quantization) be the same as for classical physics, so let us reason classically.

---

[45] Believe it or not, the factor $\frac{1}{c}$ is inserted to make the units come out right—we could have eliminated it by simply adopting units of measurement in which $c = 1$. Note, too, the difference between the magnetic potentials referred to earlier and the magnetic field. The magnetic potential does not contain any potential energy—this is all contained in the electric potential. The magnetic field, on the other hand, does produce potential energy.

If we think of the electron as orbiting classically around an atomic nucleus (as in the later abandoned "Bohr model" of the atom), then a fairly elementary calculation[46] shows that

$$(5) \qquad \gamma_{orbit} = \frac{e}{2m},$$

where e is the charge of the electron, and m its mass. And, even in quantum mechanics, the calculation is valid: though, of course, angular momentum is quantized in units of Planck's constant, the *relation* between orbital energy, angular momentum, and magnetic field remains valid—as a "quantized" operator equation.

For the "spinning" electron, understood as a top, the same classical calculation holds as for the orbiting one, yet incontestable experimental evidence yields a gyromagnetic ratio *twice* as large:[47]

$$(6) \qquad \gamma_{spin} = \frac{e}{m}.$$

Experimentally, then, the "quantized" Hamiltonian for the spinning electron in an electromagnetic field must have the term

$$(7) \qquad -\frac{e}{m}\left(\frac{1}{c}\right)\left(\frac{\hbar}{2}\,\sigma\cdot\boldsymbol{B}\right),$$

twice what we would seem to get from "quantizing" the classical formula for orbital energy. Remarkably, this very term appears in the equation derived by sorcery in Appendix C—a great victory for quantization. But why does the more direct method of quantizing yield a result which is one-half of the right answer?

Now this is not such a great anomaly as it may seem. The idea of "quantizing" electron spin is already a misnomer, since the spin of an electron is not a classical concept in the first place. That is, electrons do not literally turn on their axis; their angular momentum is not a result of spatial motions. We have also seen earlier that to return an electron to its initial position, a turn of 720 degrees (i.e., twice what we would expect) is necessary. Even what I have called the direct method of quantizing yields a result that has the right form, though

---

[46] See, for example, French and Taylor 1978, 438 ff.

[47] I ignore quantum field fluctuations which prevent this statement from being exactly true.

it has to be multiplied by 2 to get the right magnitude—that is as much as we could expect. What is astounding is that we get the right answers for the spin-coupling energy by formal factoring of energy equations—and that we get the same answer by factoring the classical energy equation as by factoring Einstein's energy equation.

* * *

My final example of a triumphant "quantization" is perhaps the most spectacular of all. I refer to the program in physics known as "gauge field theories," inaugurated by Yang and Mills in their seminal article.[48] Yang and Mills lay down a recipe for writing down quantum field equations—or, what is the same, the equation for the bosonic particles that make up the field.

As we have seen so often, the basic idea was by a brilliant, though formal, analogy to an existing theory—in this case, Maxwell's theory of electromagnetism. Yet, strangely, the analogy could not have been appreciated by Maxwell, as we shall now relate.

Maxwell's equations govern the electric and magnetic fields, including electromagnetic radiation, and show how they are generated by electric charge, which is a conserved quantity. However, physicists discovered that computations are much easier to make when, instead of two fields (with a total of six components), one *integrates* the fields to obtain potentials at each point in space—the one electric potential $\phi(\mathbf{x},t)$ and the three magnetic (or "vector") potentials $A_x(\mathbf{x},t), A_y(\mathbf{x},t), A_z(\mathbf{x},t)$. By multiplying the charge by the electric potential one gets the potential electric energy; the potential magnetic energy, as we have seen in , has a more complicated formula. The point is that one does calculations using the potentials, and then retrieves the electric field by differentiation as follows:

$$(8) \qquad E_x = -\frac{\partial}{\partial x}\phi$$

$$E_y = -\frac{\partial}{\partial y}\phi$$

$$E_z = -\frac{\partial}{\partial z}\phi,$$

---

[48] Yang and Mills 1954; Mills himself published an extremely useful retrospective article, Mills 1989. I am grateful to Issachar Unna for bringing this article to my attention.

or, for short,

(9) $$E = -\nabla\phi.$$

The magnetic field is retrieved in a more complicated way as follows:

(10) $$B_x = \frac{\partial}{\partial y} A_z - \frac{\partial}{\partial z} A_y$$

$$B_y = \frac{\partial}{\partial z} A_x - \frac{\partial}{\partial x} A_z$$

$$B_z = \frac{\partial}{\partial x} A_y - \frac{\partial}{\partial y} A_x,$$

or, for short,

(11) $$\mathbf{B} = \text{curl}(\mathbf{A}) =_{df} \nabla \times \mathbf{A}.$$

Now, whenever we integrate a function, we introduce an arbitrary "integration constant," which disappears upon redifferentiation. Since the differentiation here is only partial, we introduce an entire arbitrary function. In fact, it is not hard to show that, for any function $\theta(t,x,y,z)$, the following transformations of the potentials leave Maxwell's equations exactly as they are:

(12) $$\phi \rightarrow \phi + \frac{1}{c} \frac{\partial}{\partial t} \theta$$

$$A_x \rightarrow A_x - \frac{\partial}{\partial x} \theta$$

$$A_y \rightarrow A_y - \frac{\partial}{\partial y} \theta$$

$$A_z \rightarrow A_z - \frac{\partial}{\partial z} \theta.$$

These transformations were called "gauge" transformations, since they were thought to be trivial changes in the scale of potential energy, of no greater significance to physics than the "invariance" of the laws of nature under transformation of Fahrenheit to Celsius.

From the contemporary point of view, initiated by Hermann Weyl, however, the invariance of Maxwell's equations under gauge transformations is a symmetry, analogous to the invariance of Einstein's equations under local arbitrary coordinate transformations (general covariance). And, imitating Einstein's procedure, Weyl proved that Maxwell's equations are virtually the *only* equations that are both gauge and Lorentz invariant.[49] Weyl hoped that the formal similarity between general covariance and gauge invariance would provide the key to unifying electromagnetism with gravity, Einstein's longstanding dream.

Instead, Weyl's idea surfaced in the Yang–Mills paper of 1954 in a startling new way. In the context of quantum mechanics, charge, like any other quantity, must correspond to a symmetry—and conservation of charge must be a consequence of a physical system having that symmetry. We have seen already, however, that we cannot expect every symmetry to be a spacetime symmetry, such as spatial rotations, translations, etc. Formal, or abstract, symmetries have become the bread and butter of physics. Furthermore, we have no analogue in classical mechanics to work with—that is, we knew before quantum mechanics that linear momentum was associated with translational symmetry, that angular momentum was associated with rotational symmetry, etc. Here we must work backwards, as it were, and guess the symmetry associated with a known quantity.

Let's say that Q is the (unknown) symmetry group associated with charge and charge conservation. Suppose that $q \in Q$, i.e., q is a transformation on Hilbert space such that, if Ψ designates a system with a certain charge, so does $q\Psi$. But the truth is that charge is a quantity entirely independent of any other quantity we have discussed so far—its symmetry transformation can therefore make no physical change in the state, say, of a free charged particle. The ineluctable conclusion is: q is nothing more than multiplication by a complex number of norm 1, $e^{i\theta}$—in other words, a "phase change" of θ. Suppose the system has a charge of $-ne$. Then we can say (as a matter of simple bookkeeping) that the symmetry transformation associated with charge is multiplication by $e^{-ine\theta}$. Differentiating, we get

$$(13) \qquad \frac{\mathrm{d}}{\mathrm{d}\theta} e^{-ine\theta}|_{\theta=0} = -ine.$$

Multiplying by i (to get a self-adjoint transformation), we get the quintessentially trivial result, that the charge operator is that opera-

---

[49] An equation is *Lorentz-invariant* if it implies that the speed of light is constant in every inertial coordinate system.

tor that multiplies each and every vector in the space by $-ine$. In other words, every vector in the space has a definite charge.[50]

We have then that invariance of the laws of nature under change of phase of the state vector (an abstract change, to be sure) corresponds to charge conservation, an entirely picayune matter. The triviality of this is not alleviated by giving the group of phase changes a name: U(1). U(1) is still nothing but the group of multiplications by complex numbers of norm 1.

Things started to look very interesting, however, when Hermann Weyl pointed out that charge conservation was not simply a global phenomenon, but a local one. That is, charge moves continuously, like a fluid, from place to place—it does not "disappear" here only to "reappear" instantaneously over there. Thus invariance under change of phase should also be regarded as a local phenomenon. Namely, the transformation q must be a function of position, $q(x)$, multiplying by $e^{i\theta(x)}$, a different phase for each point in space. This, of course, must change the physical state of a state vector: we are multiplying each spatial coordinate by a different complex number (though, it is true, one which changes continuously and smoothly from point to point: $\theta(x)$ is a differentiable function of x) which can definitely change, say, the momentum of the particle as well as its energy. Where does this energy and momentum change come from?

We must therefore suppose, argued Weyl, that this momentum and energy come from a surrounding field which contains both energy and momentum in reserve, so that changes in the energy/momentum of the particle are compensated for by changes in the field.

The field is best thought of, à la Einstein, as defined by four numbers—potentials—at every point in spacetime. And, *mirabile dictu*, it turns out[51] that a *local* shift in phase $\theta(x)$ is exactly compensated for by the following shift in the potentials:

---

[50] We could, for every positive or negative integer n, make a copy of Hilbert space associated with that amount of charge. Charge is then the operator that multiplies a unit vector, if it lives in Hilbert space number n, by the number $-ne$. Conservation means that a vector trapped in one Hilbert space can never get out.

[51] See Mills 1989, 498, for details.

$$(14) \qquad A_0 \to A_0 + \frac{1}{c} \frac{\partial \theta}{\partial t}$$

$$A_1 \to A_1 - \frac{\partial \theta}{\partial x}$$

$$A_2 \to A_2 - \frac{\partial \theta}{\partial y}$$

$$A_3 \to A_3 - \frac{\partial \theta}{\partial z}.$$

But these are exactly the gauge transformations (12). And, by Weyl's other theorem, the gauge transformations (together with the Lorentz transformations of special relativity) determined Maxwell's Equations. The baffling conclusion is: the move from global to local invariance in quantum mechanics is equivalent to the existence of the classical electromagnetic field as described by Maxwell. Naturally, the electromagnetic field can itself be quantized (the field quanta are called photons) by substituting operators for variables in the classical field equations. Note, however, that the variables here are field variables; i.e., they are different at every point in spacetime. Hence, at each point in the spacetime continuum, we must follow suit in quantum field theory by substituting a different operator. Thus we have to substitute infinitely many operators in quantizing a classical field.

However astounding this result may be, it remained a mathematical theorem, derivable strictly from premises already accepted by physicists—thus giving no new information—until Yang and Mills entered the scene. One could generalize Weyl's procedure by choosing a more complicated global symmetry than U(1) symmetry, positing that the global symmetry must also be local, and writing down the appropriate equations.

Consider, as they did, the group known as SU(2), which we have come across many times in this book. These are the 2×2 complex unitary matrices of determinant unity.[52] They operate on a 2×1 array of complex numbers so as not to change the sum of the (absolute) squares of the numbers, and thus, for the mathematician, they are the natural generalization of U(1)—simple phase change.

---

[52] Recall that a unitary matrix M satisfies $MM^* = I$.

Recall that Heisenberg had conjectured in 1932 that the neutron and the proton were two different states of the same particle, just as the spin up and spin down electron are the same particle. The mathematical symmetry behind this conjecture was, indeed, SU(2) symmetry: the state vector was to be given by a two-component psi-function

$$\begin{pmatrix} \Psi_p(x) \\ \Psi_n(x) \end{pmatrix}$$

similar to the case of electron spin. A "definite" proton had a zero below; a "definite" neutron, a zero above. One could turn a proton into a neutron by transforming the coordinate function of one into the other by an SU(2) matrix. Generally, no matter what SU(2) matrix one applied, the physics was to be invariant. This symmetry was associated with the conservation of a quantity Heisenberg called "isotopic spin," by analogy to electron spin.

But here, argued Yang and Mills, the analogy ended. Isospin (as it is called today) could, and should, be regarded more like charge than like electronic angular momentum. Isospin could be considered the source of the powerful field in the nucleus, so powerful it easily overcame the repulsion of the protons. This could happen if, and only if, isospin went from a global to a local symmetry. So they calculated what kind of field it would be necessary to introduce to compensate for local SU(2) symmetry; what were, in short, the gauge transformations necessary here.

Here they discovered that an important difference between U(1) and SU(2) implied a physical difference between the electromagnetic and the (hypothetical) isospin fields. U(1) is, obviously, a commutative group—one can multiply by numbers, including complex numbers, in any order. In the case of matrices, the order of multiplication usually changes the product. This is the mathematical difference: the physical consequences of this (as explained in Mills 1989, 498–9) are that the nuclear field would itself carry isospin, unlike the electromagnetic field, which is not charged.

Rather than dwelling on this, however, let us continue in the tracks of Yang and Mills. Using the procedure discovered by Weyl, they derived the appropriate "Maxwell equations" for the nuclear field. These equations they quantized into field quanta which, in the isospin case, were charged particles (unlike photons) for reasons already referred to.

Before elucidating the improbable (for the naturalist) character of this discovery, let us make sure that Yang and Mills did make a discovery. In one sense, their paper was a failure. The Yang–Mills field quanta, like all those of gauge fields, have zero mass, and the nuclear forces required massive field quanta.[53] What is more, SU(2) symmetry is not the basic symmetry of the nucleus, as we have already seen: SU(3) is required, since the protons and neutrons are not elementary particles (they are made of quarks).

In the true sense, however, the Yang–Mills paper was one of the greatest triumphs of twentieth century physics. The problem with mass was solved by the concept of "spontaneous symmetry-breaking": massless particles can acquire mass "later." And the specific non-commutative group they chose to work with was not the main point. Other investigators soon put the Yang–Mills "algorithm" to work using other, more involved, symmetry groups. The fruits so far have been the "electroweak" theory of Glashow, Weinberg, and Salam; and the theory known as QCD for the strong forces of the nucleus. Physicists are persuaded that the ultimate theory of "everything," if it exists, will be a gauge theory. We can say, therefore, that the Yang–Mills paper was an eminent success.

Yet, there are at least three reasons why this prediction of Yang and Mills was more like "magic" than (the naturalist version of) "science."

First, the idea that "global symmetries are local symmetries" is Pythagorean thinking. The symmetries in question are abstract symmetries, i.e., not spatiotemporal symmetries, so the validity of projection from the success of one instance of this rule to another is heavily dependent upon the idea that we must categorize nature in the categories of "mathematics." One must not forget that there were only two examples of a global/local symmetry in 1954: general relativity and electromagnetism.

Second, although the electromagnetic field was a well-established

---

[53] Massless field quanta like the photon imply long range forces like the electromagnetic or gravitational forces. The nuclear forces, for example, fall off drastically with distance.

empirical phenomenon, detectable on the macroscopic level, prior to its so-called "quantization," the "classical" gauge fields of the Yang–Mills program were hardly real objects.[54] They existed only to be quantized away, for they cannot be detected at all as classical objects. The nuclear forces, for example, can be detected only at the quantum level. The idea that we write down a fictitious field equation in order to quantize it—and expect the magic to work—is even wilder than the usual procedures of quantizing things like the Lenz vector (of which I have already written that the expectation they should work is childlike). Mills himself writes:

> While gauge theories are easy to formulate at the classical level, the process of quantizing gauge theories is quite awkward, involving either noncovariant procedures or the introduction of unphysical degrees of freedom. . . . If the most basic theory of the universe is a quantum gauge theory, then a gauge theory should be the *most* natural thing (if not perhaps the *only* thing) that can be quantized, rather than the most awkward; indeed, you should be able to formulate a quantum gauge theory directly, without going through the intermediate stage of the classical theory. (Mills 1989, 507)

It is interesting, by the way, that the equation that Yang and Mills wrote down, as a generalization of Maxwell's equation, could be seen by them *as* a generalization only via the formalism of quantum mechanics. The analogy between the two equations—one set linear (Maxwell's) and the other nonlinear—was not a *direct* mathematical analogy.[55]

Finally, when we look at the quantization procedures themselves, we are confronted with a difficulty going well beyond those of previous examples of quantization. The quantization "rules" involve, as I have said, the introduction of a continuum of operators to replace field variables in (pseudo) "classical" equations. These equations have the form

---

[54] I mean "physically real." Mathematically, the concepts of gauge field theory are isomorphic to those of the *geometrical* theory of "fiber bundles." See Chapter 2 for a discussion.

[55] Of course, given the geometrical apparatus of fiber bundles, one can formulate the analogy without quantum mechanics. But, as I have argued in Chapter 2, Yang and Mills did not see this analogy at all.

$$(15) \qquad\qquad P(x)Q(x) - Q(x)P(x) = \ldots,$$

where we must solve for $P(x)$ and $Q(x)$. The discouraging fact is, that to this very day it is not known whether equation(s) are consistent or not—whether they have solutions in some kind of reasonable "mathematical" structure (as did the square root of minus one ultimately) or not. Certainly, Yang and Mills had no evidence that these equations had solutions.

Physicists know how to manipulate equations to get numbers without solving them, however, and they have thus been able to make predictions from gauge field theories without solving the question of mathematical consistency. The success of these predictions is all the more startling if we do not realize that it is the formalism itself (and not what it means) that is the fundamental subject of physics today.

In summation: the most startling success of quantization, namely, the Yang–Mills procedure, involved "quantizing" a fictitious "classical" equation C, arriving at a quantum field equation Q not known even to be consistent. The classical equation itself, C, was arrived at by an analogy to other classical field equations—a formalist analogy which was (to Yang and Mills) incomprehensible without the formalism of quantum mechanics.[56]

* * *

The story of quantization, in sum, buttresses the major theses of this book. The early founders of quantum mechanics, particularly Bohr, spoke of a "correspondence principle" relating classical and quantum mechanics. But I have shown here, I think, that this correspondence principle was deeply anthropocentric (because formalist). Hence, the true "correspondence" was between the human brain and the physical world as a whole. The world, in other words, *looks* "user friendly." This is a challenge to naturalism.

---

[56] I emphasize that this formalist analogy was "upgraded" to a Pythagorean analogy later on, when the theory of fiber bundles provided a direct analogy between the various classical gauge field equations.

Appendix A

A "Nonphysical" Derivation of Quantum Mechanics

There are many textbooks on quantum mechanics, and I do not aim to increase their number. My goal here is to show, in greater detail than in the text, the power of the Hilbert space concept—its descriptive applicability to quantum mechanics. I will do this by showing how the formalism of quantum mechanics actually suggests—nondeductively—its own development. That is, we can "read off" information about the world—in a "nonphysical," yet nondeductive way—from the formalism. In this way we will get a feel for the mysterious nature of the Hilbert space concept, in its application to physics.[1]

There is nothing historical, therefore, about the following "derivation" of quantum mechanics, which was inspired by, and resembles, that of Henry 1990.[2] On the contrary, I reverse the historical order, to show that, starting with little more than the "Maximality Principle," quantum mechanics could have been discovered by studying the formalism itself, rather than studying nature. (Later, I will discuss some actual discoveries using "formalist" reasoning.) This

---

[1] I emphasize once again that the mystery could well be solved in the future. I am writing from the point of view of present knowledge.

[2] The present text, however, incorporates improvements I learned from Joel Gersten, Percy Deift, Shmuel Elitzur, Sylvain Cappell, Shelly Goldstein, and Harry Furstenberg.

Henry's aim was completely different from mine. His treatment was meant for the classroom, to persuade students that quantum mechanics is "inevitable." Needless to say, I dissociate myself from that goal.

certainly shows how remarkably suited is the Hilbert space concept to describe nature.

Suppose, provisionally, that a particle can be in one of n places: $x_1, x_2, \ldots, x_n$, and that for each we know the probability that the particle is at x_i. The "state" of the particle can be represented by a unit vector in an n-dimensional space, one "axis" for each possible position of the particle, where:

A. The n coordinate axes are perpendicular to one another;

B. The *square* of the coordinate of the unit vector with respect to "axis x_i" gives the probability of the particle's being at place x_i. (The sum of the squares of the coordinates of the vector equals 1.) We use the squares of the coordinates so that the vector's length always equals 1—a convenience (in principle, we could allow the coordinates themselves to be the probabilities, and then the sum of the coordinates would equal 1).

So far, no *physics* has been done.

On the same n-dimensional space, superimpose a new set of perpendicular axes, $p_1, p_2, \ldots, p_n$. The numbers $p_1, p_2, \ldots, p_n$ represent the possible momenta of the particle, and the new axes are placed so that the new coordinates, squared, give the probability that the particle has the corresponding momentum.

In the real world, however, there are as many places a particle could be as the power of the continuum—and so for momenta. Thus our space needs to have axes equal in number to the power of the continuum, both for position and for momentum. Instead of saying that the sum of the squares of the coordinates is 1, we now say that the integral of the squares of the coordinates is 1.[3] That is, if Ψ is the state vector, so that the coordinates of the vector are given by the expression $\Psi(x)$ (the coordinate of the vector at "axis" x), the "length" of the vector is defined as $\int_{-\infty}^{+\infty} |\Psi(x)|^2 dx$, and a unit vector is defined by the equation $\int_{-\infty}^{+\infty} |\Psi(x)|^2 dx = 1$.

Here, formidable problems arise—the analogy between finite- and infinte-dimensional spaces is tricky. Most vectors, according to this

---

[3] Of course, we are now interpreting the squares of the coordinates as probability *densities* (probability per centimeter, say) rather than as simple probabilities. For ease of reading I shall often write sums when I really mean integrals.

definition, have no "length" (norm), because the integral is infinite. This will happen not only if the absolute value of the function, $|\Psi(x)|$, is unbounded, but even if it is constant.

The "converse" problem also arises: Because we would like to have something as a unit vector, its coordinates are unbounded and go to infinity. Think of a particle which is with probability 1, i.e., with certainty, at position x_0. This means that the function $\Psi(x)$ is zero whenever $x \neq x_0$. But if $\Psi(x_0)$ is finite, the "length" of the vector is also zero (the area under a curve cannot change by changing the curve on a finite set of points, or in general on a "set of measure zero").

The space of vectors Ψ with a defined, i.e., finite, length—where we identify any two vectors whose coordinates are "almost all" the same[4]—is the classic Hilbert space, known by mathematicians today as L2. In general, any vector space on which we can define the length of vectors, or more generally the length of the projection of one vector on the other, determines a Hilbert space; but in what follows, "Hilbert space" will mean the classic one. However, following the physicists, and throwing rigor to the winds, we shall speak as though there were in the Hilbert space a vector all of whose coordinates are equal, and also a vector of length 1, whose coordinates are "almost all" zero. The reason these vectors are useful to physicists is that both can be thought of as infinite limits of vectors that are in the Hilbert space. For example, think of a bell curve around a point; you can make the curve as narrow around the point as you want without changing the area under the curve, so long as you make it higher and higher. Similarly, you can pull a bell curve apart without limit without changing its area, because you can make it lower and lower. Hence these two kinds of vectors can be thought of as ideal elements, to be added to the Hilbert space. (The mathematicians have already thought of ways to make this rigorous.) In the following discussion, I will therefore speak as though these ideal elements were part of Hilbert space. In any case, my view is that major discoveries were made by such "incorrect" mathematics—i.e., where the formalism had not been interpreted consistently.

* * *

[4] Technically, this means that the members of Hilbert spaces are equivalence classes of vectors, but we won't bother much with this distinction.

The requirement that a *single* unit vector give us information *both* about the position (probabilities) of the particle *and*—by a change of basis—about the momentum (probabilities) of the particle is an example of what I have called the *Maximality Principle*. Unfortunately, calling it a "principle" does not make it true that one vector *could* give all this information, in this way.[5] For example, suppose we measure position of the particle with respect to a measuring device, which we then move. The position coordinates of the particle "shift" to reflect the relocation—the state vector moves ("rotates") *in the Hilbert space*. Yet the *momentum* probabilities should not change just because we measure them from somewhere else. Relative to the momentum coordinates, the state vector should not "move" at all. But how could it move relative to only one set of axes?

Consider a particle in one *spatial* dimension (that is, confined to the x-axis) whose position and momentum probabilities are given by the (squares of) the coordinates of a single state vector Ψ.[6] Write $\Psi^{pos}(x)$ for the coordinates of the vector with respect to the "position axes," and $\Psi^{mom}(p)$ for the coordinates of the vector with respect to the "momentum axes." Thus the probability that the particle is at position x is $(\Psi^{pos}(x))^2$, and the probability that the particle has momentum p_x is $(\Psi^{mom}(p))^2$. The graph of both functions might be a bell curve. In what follows, I will be assuming $\Psi(x)$ to be differentiable.

Now suppose we move our measuring instrument as before. In fact, to apply the calculus, assume that it is moved an "infinitesimal" amount ε,[7] say in the "negative" direction. The state vector should shift in *its* space, so that its new coordinates are given by the function $\Psi_{new}(x) = \Psi(x + \varepsilon)$. The "bell curve" of the probability will have the same shape, but shifted the amount ε in the "positive" direction.

---

[5] Whose importance Shelly Goldstein in particular impressed upon me. That Henry 1990 glides over the issue weakens the credibility of his claim that quantum mechanics is "inevitable."

[6] It is crucial to keep in mind the distinction between physical space and Hilbert space. The physical particle is in physical space—which we are restricting to the x-axis for simplicity. The Hilbert space is infinite-dimensional.

[7] The free use of the intuitive "infinitesimal" notation by physicists was universally considered mathematically inconsistent until Abraham Robinson settled the matter by proving that it is, in fact, consistent to assume the existence of numbers, though not being equal to zero, are yet smaller than any real numbers.

Since the vector has shifted, albeit by an infinitesimal amount, the momentum coordinates $\Psi^{mom}(p)$ necessarily shift as well. How could they change, without changing the momentum probabilities? Only, it would seem, by multiplying the coordinates by ±1; that is, some of them by $+1$ (which is too little, i.e,. no change at all) and some of by them by -1 (which is too much, i.e., not an infinitesimal change). And, recall, any other change would alter the momentum probabilities.

To preserve maximality—to represent both position *and* momentum with the same state vector—we must find numbers that are infinitesimally close to 1, but have absolute value 1. The complex numbers can provide what we need.[8] The "complex" numbers are two-dimensional vectors (not to be confused with our state vector)—they have both "length" and "direction." The number $+1$ gets reinterpreted as a vector along the "real axis" having directional angle zero; $+i$, the square root of -1, is a vector of length 1 with direction 90° "north"; etc. We represent a vector of length 1 and direction angle δ as

$$e^{i\delta},$$

and if δ is infinitesimal, we have just what we want: a number infinitesimally different from $+1$, but whose absolute value (length) is 1. Thus, if our state vector moves infinitesimally, we can multiply each momentum coordinate $\Psi^{mom}(p)$ by a complex number whose direction differs infinitesimally from 0°. The result is now a complex number $e^{i\delta_p}\Psi(p)$ whose *absolute*[9] square is still $|\Psi^{mom}(p)|$.

So the Maximality Principle forces us to introduce complex coordinates.[10] The complex coordinates of a state vector with respect to the position "axes" are called the *amplitudes* of the position. We need

[8] There are other mathematical structures that have been called "numbers"—quaternions, octonions, "p-adic" numbers, and others. One could imagine quantum mechanics done with these algebras. The complex numbers, nevertheless, allow the most straightforward extension of the concepts of analysis (the calculus).

[9] We must be careful here: the square of a complex number is itself a complex number. We take here the *absolute* square: the square of the length of the vector, which is its "absolute value." For "real" numbers there is no difference between the square and the absolute square, because the square of a negative number "happens" to be a positive.

[10] Strictly speaking, what is forced is the abandonment of the real number field for a higher structure. For this comment I thank Sylvain Cappell.

a new word, because the coordinates do not give (even) the probabilities of position, except through their absolute squares. The probabilities, of course, are still real: they are the absolute squares of the coordinates.

In classical mechanics, the positions and momenta of particles are independent of one another. Yet in the quantum formalism, with the Maximality Principle, if we know the coordinates of the state vector with respect to the position axes, we can calculate them with respect to the momentum axes—a truly remarkable fact. In other words, the Maximality Principle fixes inalterably the "orientation" of the momentum axes with respect to the position axes. This is true whether or not momentum is conserved.

To see this, consider a state vector Ψ_p lying *along* one of the momentum axes—representing, therefore, with certainty (probability 1) a particle of momentum p.[11] So the coordinate, *squared*, of Ψ_p, with respect to the p-momentum axis, is unity: in symbols, $|\Psi_p(x)|^2 = 1$. (The rest of the coordinates are zero.) Let us see how vector Ψ_p is changed by an infinitesimal translation of our measuring device by ε in the negative direction.

We have, by our previous argument,

$$(1) \qquad \Psi_p^{pos}(x + \varepsilon) = e^{i\delta}\Psi_p^{pos}(x).^{12}$$

The phase δ can obviously depend on the momentum p and the amount of the translation ε. But we can *require* that δ be independent of the position x (that is, we require that the function $\Psi_p(x)$ be such that δ be independent of x). This is the simplest postulate; it also reflects our belief that the points of space are indistinguishable. Let us see what this postulate yields.

---

[11] The astute reader will realize that we are speaking here of an infinite probability density, since all the probability from the entire continuum of momenta is concentrated at a point. (I warned above in the text that we would be discussing such things.) Physicists usually ignore mathematical contradictions connected with infinity, hoping that the mathematicians will give content to their meaningless formalisms. In this particular case, a mathematically rigorous way was discovered to make sense of these densities, just as a rigorous way was discovered to describe Newtonian "mass points" that also have infinite density.

[12] I shall begin dropping the superscript "*pos*" without further notice.

The left side of this equation is approximately equal to

(2) $$\Psi_p(x) + \varepsilon \frac{d}{dx}\Psi_p(x);$$

while the right side is approximately equal to

(3) $$(1 + i\delta)\Psi_p(x);$$

so we can write, to "first order,"

(4) $$\Psi_p(x) + \varepsilon \frac{d}{dx}\Psi_p(x) = (1 + i\delta)\Psi_p;$$

collecting terms, we get

(5) $$-\varepsilon \frac{d}{dx}\Psi_p(x) = i\delta\Psi_p(x),$$

that is,

(6) $$\frac{d}{dx}\Psi_p(x) = i\frac{\delta}{\varepsilon}\Psi_p(x).$$

Now the quotient $\frac{\delta}{\varepsilon}$ of the two infinitesimals must be a single real number, because we have set δ independent of x. This, of course, was a condition, ultimately, on the function $\Psi_p(x)$—justified by the "isotropy" of space. Equation (6) makes it clear that, by multiplying ("rescaling") the function $\Psi_p(x)$ by a suitable "fudge factor," we can make the quotient $\frac{\delta}{\varepsilon}$ be any real number we want. For bookkeeping only, we set

(7) $$\frac{\delta}{\varepsilon} = p;$$

the resulting differential equation is

(8) $$\frac{d}{dx}\Psi_p(x) = ip\Psi_p(x),$$

which can also be written in the illuminating form

(9) $$\left\{-i\frac{d}{dx}\right\}\Psi_p(x) = \Psi_p(x),$$

which states that the operator $-i\dfrac{d}{dx}$ "extracts" the value of the momentum, p, from the (unit vector lying in) the momentum axis.[13] Of course, some of this is mere bookkeeping, since we could have chosen any real number for the value of $\dfrac{\delta}{\varepsilon}$ in equation (7). What is not bookkeeping is the fact that the operator

$$\left\{-i\frac{d}{dx}\right\}\Psi_p(x) = \Psi_p(x)$$

acts on the unit vector by multiplying it by a number.

The solution of the equation is:

(10) $$\Psi_p(x) = ce^{ipx},$$

where c is a complex constant (it doesn't really matter which). For every x, this is a complex number whose length is that of c, where the "direction" is determined by the momentum, p. So the probability for a particle, having definite momentum p, to be at any given place x, is

(11) $$|ce^{ipx}|^2 = |c|^2,$$

which means that the particle could be anywhere, with equal probability.[14] This is, of course, a special case of the Heisenberg Uncertainty Principle.

(Actually, we should have expected this result already: if the momentum, and the momentum probability, does not change merely because of a translation of the measuring apparatus, then the probability curve remains constant when it is shifted along the x-axis. Only a horizontal straight line has this property.)

[13] These equations show that the operator d/dx plays a double role: it represents infinitesimal translations; multiplied by $-i$ (whence it becomes a Hermitian operator), it represents that quantity whose conservation follows from invariance under infinitesimal translations. See Guillemin and Sternberg 1990b for an illuminating account of the "dual role" of symmetry operators in quantum mechanics.

[14] Literally, the formula means nothing, because since the probabilities are equal, and there are infinitely many of them, they do not add up to 1:

$$\int_{-\infty}^{+\infty} |ce^{ipx}|^2\,dx = \infty.$$

Physicists ignored this problem too, relying on the mathematicians to solve it.

We can think of ce^{ipx}, as c varies, as determining the entire momentum "axis" corresponding to momentum p (with respect to the position axes)—thus, we have discovered "where the momentum axis is," if this formalism is to make sense. Where are the other momentum axes? They must be perpendicular (orthogonal) to this one. Mathematics to the rescue: any two vectors having the coordinates e^{ip_1x} and e^{ip_2x} are perpendicular:

(12) $$\int_{-\infty}^{+\infty} e^{ip_1x}e^{ip_2x}dx = 0, p_1 \neq p_2.$$

Thus, there is nothing inconsistent with adopting the "bookkeeping" device of "labeling" each momentum axis with the value assigned to it.[15] The operator $P = -i\dfrac{d}{dx}$ is called the "momentum operator," and a vector Ψ_p lying along the momentum axis satisfies

(13) $$P\Psi_p = p\Psi_p,$$

that is, the effect of the operator P on this vector is simply that of multiplying it by the real number p. Thus Ψ_p is called an "eigenvector" of P, and p is called its "eigenvalue."

Let us sum up the results so far: from little more than the Maximality Principle (a strong, yet formal, requirement) and a trivial property of momentum (invariance under translation), we have shown how momentum information must depend upon position information. The argument holds whether or not momentum is conserved. That is, the momentum axes are fixed with respect to the position axes—independently of the forces operating on the particle. The argument was not deductive, because our assumptions had to do not only with truth, but with meaningfulness. We were "forced" to introduce complex numbers, to save the meaningfulness of the Hilbert space formalism.

But the use of complex numbers has profound implications. Every coordinate will have, from now on, not only an absolute value, but also a "phase." The phases of the coordinates are what cause par-

---

[15] In other words, the momentum axis must be a complex exponential of the form e^{icx}; the assignment $c = p$ is a convention.

ticles to behave, mathematically, like waves. All the surprising features of quantum mechanics that we shall examine rest upon these "innocent" phases that leave the probabilities untouched (because they disappear upon taking the absolute square).

To repeat, this argument for the indispensability of complex numbers is retrospective.[16] Schroedinger's introduction of complex numbers into quantum mechanics was quite different (he was not working in the framework of an abstract vector space). Nevertheless, it, too, was formal.[17]

So far, then, our formalism for keeping track of probabilities, using a unit vector in a multidimensional space to describe the state, based on the Maximality Principle, has led us to two remarkable, but nondeductive, conclusions:[18]

(a) that everything knowable about the momentum of a particle can be inferred from the position (function) of the particle, irrespective of what experiment we are running; and

(b) that the position function of a particle must be complex-valued, a fancy way of saying that a particle behaves like a wave.

An even more remarkable conclusion surfaces when we apply this formalism to angular momentum.[19] Suppose the particle moves in a

[16] We see here the parallel development of physics and pure mathematics: the imaginary numbers themselves were a prime example of extending mathematics by formal manipulations. The Italian mathematician Cardano introduced imaginary numbers formally, as solutions of problems like: find two numbers whose sum is 10 and whose product is 40. Mark Wilson (Wilson 1992) recounts how nineteenth-century mathematicians used imaginary numbers to "find" the "intersection points" of a circle and a line which, on the Euclidean plane, don't meet at all. This formal move led to advances in projective geometry: e.g., the unified treatment of geometrical properties that heretofore had not seemed associated. It took hundreds of years for physics to introduce complex numbers in the "indispensable" way that Schroedinger did.

[17] See Chapter 4 for a detailed description of Schroedinger's derivation of the wave equation, including the introduction of complex wave functions.

[18] The conclusions are not deductive, because we reason, not from the truth, but from the meaningfulness, of the premises.

[19] For two-dimensional circular motion around the origin, the angular momentum is given by multiplying the (absolute value of the) momentum of the particle by its distance from the origin. For clockwise motion, we assign positive angular momentum; for counter-clockwise, negative. For other motion, we draw a radius from the origin to the particle, and then consider only the component of the momentum perpendicular to the radius.

plane,[20] so that it has two *spatial* coordinates, and one component of angular momentum. It is most convenient to take these as polar coordinates. We still have an infinite-dimensional space with an axis for each point in the place; but the function Ψ that gives the coordinates of the state vector becomes a function $\Psi(r,\phi)$ of two variables. Thus, each "position axis" is an "r–ϕ" axis, and $|\Psi(r,\phi)|^2$ is the probability (density) that the particle is at position $<r,\phi>$.

Suppose we now superimpose "axes" of angular momentum on our space—and extend the Maximality Principle to angular momentum. That is, the same unit vector that describes position "amplitudes" should describe those of angular momentum as well. Then a trifling symmetry argument, using the invariance of angular momentum under *rotation,* fixes the angular momentum "axes," as follows.

If we rotate our measuring device by an infinitesimal angle, the state vector of the system must have almost the same angular momentum coordinates as before (i.e., their absolute squares remain the same), except that the coordinates can—and must—be multiplied by a complex number with an infinitely small "angle" (remember to distinguish between the angle of a complex number and the angle of our measuring device in space). In analogy to the case of linear momentum, writing l for angular momentum and $\Psi_l(r,\phi)$ for the coordinates with respect to the "polar position axes" of a unit vector which describes a particle with definite angular momentum l, though not necessarily with definite radius, we arrive at the equation

$$(14) \qquad -i\frac{\partial}{\partial\phi}\Psi_l(r,\phi) = l\Psi_l(r,\phi),$$

whose solution is

$$(15) \qquad \Psi_l(r,\phi) = f(r)e^{il\phi},$$

for $f(r)$ an arbitrary complex-valued function. This function $f(r)$, by

---

[20] In the "real world," no particle can be restricted to a plane, because, supposing the particle moving in the xy plane, both the position and the momentum of the particle in the z-direction (zero) would be known with certainty, and we have shown that this is impossible—this was one of the mistakes of the "Bohr model" of the atom. (This point will assume great significance later on.) Nevertheless, our results below will hold true for the *z-coordinate* of angular momentum even in the "real world."

the way, can be one whose absolute square gives an immense probability density for a certain fixed distance r_0 from the origin, so that if we fudge the difference between "immense" and "infinite" (as physicists are wont to do), we can say that there is no contradiction between a particle having a determinate angular momentum and a determinate radial position. Thus, we can speak of r–l axes instead of r–ϕ axes, l replacing ϕ—a state vector lying on one of these axes having, with probability 1, radius r, and angular momentum l.

For a fixed r, though, the relationship between l and ϕ is analogous to that between p and x—the mathematics is almost precisely the same. Namely, for a fixed radius r, if the angular momentum l is also fixed, then we lose information about the angle ϕ (the orientation of the particle in its plane of motion); for

$$(16) \qquad |\Psi_l(r,\phi)|^2 = |f(r)e^{il\phi}|^2 = |f(r)|^2,$$

i.e., it is the same for all ϕ.

So far, angular momentum is a replay of linear momentum. This is not surprising: there are many analogies in classical physics between linear and circular motion. But now comes the "voodoo."

* * *

In modern logic, the word "function" denotes what is called by mathematicians, pleonastically, a "single-valued function." Functions are defined *extensionally*: there is no difference between a function and its graph: a function is a set of ordered pairs $<x,y>$, such that for each x there is exactly one y.

An older view (which survives in some complex analysis texts and in physicists' jargon) is that a function is a *rule*, and in any case, is not equivalent to its graph. According to this view, it is not a tautology that functions are single-valued. In fact, there are functions that are *intrinsically* multiple-valued.

An example is the square root function on the *complex* plane. If a complex number is a vector $<r,\theta>$ (r is called the *modulus* and θ the *phase*), its square root is the vector

$$\left\langle \sqrt{(r)}, \frac{\theta}{2} \right\rangle:$$

take the square root of the modulus and halve the phase. For simplicity, consider the complex numbers with modulus one (so the modulus will not change under square root). Begin with the vector at phase zero, and rotate it continuously counter-clockwise through 2π (360°). At the starting point, when the phase is zero, representing the real number 1, the square root is naturally the same vector, $\sqrt{1} = 1$. When the phase reaches 2π (360°), the square root is -1—i.e., a unit vector with phase π. Yet a turn of 360 degrees brings the vector back to the starting point, so that, as a result of a continuous motion, the square root function has arrived at a different value: -1. (Of course, everybody knows that 1, as an integer or as a real number, has two square roots, ± 1; the point is that in the complex plane it is possible to go from one value to the other by a continuous path.)

What about a function $\Psi(x,y)$ or $\Psi(r,\phi)$, the function that gives the position coordinates of our state vector? Must it be single-valued? You might say that, physically, a particle ought to have one position—but the only information the state vector gives is position *probabilities*. And even these it only gives by way of the absolute square, $|\Psi(r,\phi)|^2$. This means that we could multiply the function $\Psi(r,\phi)$ by some phase factor—i.e., a complex number $e^{i\theta}$ of modulus 1—without changing the physical meaning of the function.

Nevertheless, there is good reason to require that the position function be single-valued.[21] For if the position function is continuous and defined at every point $<x,y>$ in the xy plane, yet multiple-valued, we could find a loop beginning and ending at $<x,y>$ such that the value of the function at the beginning of the loop would be different from that at the same point at the end. That loop could then be shrunk continuously to infinitesimal size, so that the same—finite—change of phase would occur at an infinitesimal interval, contradicting continuity.[22]

Let us return to a function $f(r)e^{il\phi}$, which gives the amplitude (coordinate) of the direction of the particle when the angular

[21] The appropriate language for this paragraph would be that of "covering space theory" in topology. My apologies to professional topologists.

[22] With the square root function itself, this does not happen, because, as we shrink the loop around the origin to radius zero, the "distance" between the two values of the square root also goes to zero; that is, the square root function is *single*-valued at zero, double-valued everywhere else.

momentum is determined to be *l*—and let us therefore make the assumption that this function is to be single-valued, and, for simplicity, let us make $f(r)$ a constant function: the particle has equal probability to be any distance from the origin. Now the angles ϕ, $\phi\pm2\pi$, $\phi\pm4\pi$, . . . all represent the same direction; so for single-valuedness we need

(17) $$e^{il\phi} = e^{il(\phi + 2\pi)} = \ldots .$$

But this is impossible unless *l* is an integer—the number of units of the fundamental angular momentum.[23] It is customary to name this natural unit of angular momentum $\hbar$, Planck's constant. We have arrived at quantization of angular momentum, one of the fundamental empirical discoveries of twentieth-century mechanics, "for free"! The main premises here were the Maximality Principle and the single-valuedness of the coordinate function Ψ. Again, the formalism seems to display information beyond that for which it is designed. Our conclusion, the quantization of angular momentum, is the only one which allows us coherently to express data about this magnitude using a unit vector in a linear space.

Is the preceding an *a priori* argument? Of course not: you cannot arrive at empirical conclusions without making at least some empirical assumptions. (This truism is what is correct about "empiricism.")[24] But the assumptions made here are about the empirical adequacy of a formalism, not about causal processes.

---

[23] The argument I have given is valid for "orbital" angular momentum—that angular momentum that arises from physical movement of a particle around a point chosen as the "origin." As we shall see, and as knowledgeable readers already know, there is also a form of "intrinsic" angular momentum, called "spin," that can be attributed to certain elementary particles, such as the electron. The spin of the electron is ½ Planck's constant.

Nevertheless, spin ½ can exist only in three dimensions (for reasons which are far from obvious and which I shall give later). Our argument, which assumes that space is a Euclidean plane, is technically valid—in a plane, there is no angular momentum less than Planck's constant. As the reader will see by continuing this Appendix, it is quite worthwhile to explore what quantum mechanics looks like in two dimensions, as compared to three.

Even in three dimensions, though the maximality principle fails to predict half-integer spin, there are other considerations, just as "formal," which do predict it. More of this later.

[24] Nevertheless, philosophers often content themselves with this observation, over-

* * *

Next, I show the intrinsic spin of the electron, which is ½ Planck's constant,[25] though not representable by a unit vector on our linear space, is nevertheless hinted at by the quantum formalism more generally.

In our treatment of angular momentum, we discovered the "location" of the axes corresponding to various values of angular momentum by a symmetry argument: angular momentum was that quantity which stays the same under rotation. A further argument, based on the single-valuedness of the coordinate function, yielded the result that the possible values of angular momentum are integral multiples of a minimum value, Planck's constant.

This treatment, however, based as it was on two-dimensional motion, is oversimplified. Adding the third dimension makes for quite a difference. I shall now sketch what happens in three dimensional motion, leaving out most of the mathematical details and all of the proofs.

Any rotation in three dimensions is the sum of rotations around, successively, the x, the y, and the z axes. Instead of speaking of angular momentum, we must speak of the three components of angular momentum. Furthermore, it is no longer true that angular momentum is invariant under rotations: the angular momentum vector will change direction, if the rotation is not around the vector itself. What remains true, however, is this: any rotation around the x, the y, and the z axes will not change the corresponding component of the angular momentum vector. This is *a fortiori* true for infinitesimal rotations, and so we get three operators, L_x, L_y, L_z, for each of the components of angular momentum, corresponding to the three infinitesimal rotation operators

---

looking that there can be *degrees* of *a prioricity*. What is surprising in physics is how much can be obtained from so little.

[25] If we think of angular momentum as a three dimensional vector, according to the classical picture, then "spin ½" does not mean that the length of the vector is ½ Planck's constant, but rather that the maximum possible value of the z coordinate of the "vector" is ½. The validity of the vectorial picture here is limited in any case, because (as we shall see) it is impossible to determine more than one of the components of angular momentum exactly, and therefore we cannot use the Pythagorean Theorem to calculate the length of the angular momentum "vector" in the standard way.

(18)
$$J_x = -i\hbar L_x,$$
$$J_y = -i\hbar L_y,$$
$$J_z = -i\hbar L_z.$$

As before, the three angular momentum operators L_x, L_y, L_z (via the eigenvector equations) determine the position of the "axes" assigned to each value of angular momentum. Yet here an extraordinary contrast with linear momentum arises: the axes for the three components of angular momentum can never coincide. The reason for this, mathematically, is that the three angular momentum operators do not commute with one another, and thus cannot have common eigenvectors. In other words, should we know the exact value of the z-coordinate of angular momentum, the other two coordinates are absolutely uncertain—in stark contrast to the case of linear momentum.

Mathematically, the formula connecting three operators of angular momentum is:

(19)
$$L_x L_y - L_y L_x = i\hbar L_z$$
$$L_x L_z - L_z L_x = i\hbar L_y$$
$$L_y L_z - L_z L_y = i\hbar L_x.$$

From now on, when we think of a particle with some fixed angular momentum, we will mean one with a fixed z-component of angular momentum (which is all that we can mean). As before, though, this fixed angular momentum is quantized: it must be some integral multiple of Planck's constant. Suppose, then, the angular momentum is 5 units—meaning the projection of the angular momentum on the z-axis is 5 units. Since we do not know the other two components of the angular momentum vector, the value of the z-component (namely 5) gives us less information than we might expect. If we rotated our laboratory away from the vertical, so that the new z-axis no longer is perpendicular to the floor, the (z-component of) the angular momentum would change: for example, it might increase, say, to 7 units. In fact, to get maximal information about the (z-component of) angular momentum, one needs to know not only the value, but what values it could have if the laboratory were rotated away from the vertical. Actually, it would be enough to know the *maximum* value that the angular momentum could take on. For example, by rotating the laboratory away from the vertical by a certain angle, the (z-component of) angular momentum might change from 5 to 9 units and then begin to decrease again. Or, it might increase to 1,000,000 units. (In the classical picture, this would happen if the angular momentum "vector" was very long but almost horizontal in position, so that its "shadow" on the z-axis was very small.) On the other hand, 5 might be the maximum value under such a rotation. So, then, the angular momentum axes must contain information of the form (l,m), where l is the (z-coordinate of) angular momentum and m is the maximum possible under rotation. Clearly, then, if m is the maximum possible angular momentum for a system, then $-m$ would be the minimum. But what are the values in between?

In order to solve this problem, consider the problem in terms of the state vector in our linear space, rather than in terms of physical space. As we rotate our laboratory in all directions, the state vector moves too. But if it happens to be located in a subspace determined by the permitted values (l,m) for a fixed m, it will move around in that subspace, trapped. Note that this subspace must be of *finite* dimension; for there are only a finite number of values of angular momentum possible where the maximum value is fixed. When restricted to a finite linear space, though, the angular momentum operators L_x, L_y, L_z reduce to self-adjoint matrices M_x, M_y, M_z. The fact that the matrices do not commute means, algebraically, that we cannot diagonalize the matrices simultaneously, but only one of them at a time; naturally, we diagonalize M_z. The diagonal of that matrix gives us the values of permitted (z-coordinate of) angular momentum for a fixed maximum.

The problem has now been converted to an equation, or rather three equations. We are looking for three Hermitian matrices, M_x, M_y, M_z, one of which (M_z) is diagonal, that satisfy the following equations:

(20)
$$M_x M_y - M_y M_x = i\hbar M_z$$
$$M_x M_z - M_z M_x = i\hbar M_y$$
$$M_y M_z - M_z M_y = i\hbar M_x.$$

Solving these equations is a piece of algebra,[26] which yields the following result: the matrix M_z necessarily has the form

$$(21) \quad \begin{pmatrix} m & 0 & 0 & . & 0 & 0 \\ 0 & m-1 & 0 & . & 0 & 0 \\ 0 & 0 & m-2 & . & 0 & 0 \\ . & . & . & . & . & . \\ 0 & 0 & 0 & . & -m+1 & 0 \\ 0 & 0 & 0 & . & 0 & -m \end{pmatrix}$$

where m may be an integer or a half-integer (i.e., an integer plus ½ such as ½, 1½, 2½, etc.).

Thus, for example, a particle may be in a state where the z-coordinate of its angular momentum may be one of the following: 3, 2, 1, 0, −1, −2, −3 (times Planck's constant). Tilting the laboratory away from the vertical by suitable angles will "change" this value to one of the others—and one of the others only. If we turn ourselves upside down, of course, the value will assume its negative.

But what of the half-angular values? We have already demonstrated that angular momentum must come in integral multiples of Planck's constant. Only thus can the "wave function" (for us, the coordinate function) be single-valued. Consider, for example, the case of "angular momentum ½." The equations are satisfied by the following three matrices, of which only the last is diagonal:

$$(22) \quad M_x = \begin{pmatrix} 0 & \frac{1}{2}\hbar \\ \frac{1}{2}\hbar & 0 \end{pmatrix}, M_y = \begin{pmatrix} 0 & \frac{1}{2}\hbar i \\ -\frac{1}{2}\hbar i & 0 \end{pmatrix}, M_z = \begin{pmatrix} \frac{1}{2}\hbar & 0 \\ 0 & -\frac{1}{2}\hbar \end{pmatrix}.$$

Defining the three "Pauli spin matrices" by

$$(23) \quad \sigma_x = \begin{pmatrix} 0 & 1 \\ 1 & 0 \end{pmatrix}, \sigma_y = \begin{pmatrix} 0 & i \\ -i & 0 \end{pmatrix}, \sigma_z = \begin{pmatrix} 1 & 0 \\ 0 & -1 \end{pmatrix},$$

we can write the solutions as follows:

$$(24) \quad M_i = \frac{1}{2}\hbar\sigma_i, i = 1, 2, 3.$$

[26] For a nice exposition, see Yamanouchi 1970.

Equation (22), particularly the form of Mz, seems to suggest that there could be a particle whose (z-coordinate of) momentum is $\pm\frac{1}{2}$. But this leads to a paradox: the function of a particle with (z-coordinate of) angular momentum $\frac{1}{2}$ would have to be some function of the radius of the particle multiplied by

(25) $e^{i\frac{\hbar}{2}\phi}$.

And such a function is double-valued: to put it another way, to make this function return to its initial value, one must rotate the z-axis not 360°, but 720°.

There would be no logical objection simply to *rejecting* the half-angle values of angular momentum as "unphysical." After all, the equations set down necessary conditions for the matrices we seek. Perhaps they are not sufficient. It is not unknown for equations in physics to have solutions that are rejected—perhaps the most famous example being Einstein's rejection of velocities faster than the speed of light.

But, on the other hand, mathematical possibilities have turned out to be physically real far more often than might have been expected. And spin $\frac{1}{2}$, though bizarre indeed, turns out to be an empirical reality. Every electron turned out to have an intrinsic angular momentum of $\frac{1}{2}$ Planck's constant, as if it were spinning on an axis, and thus can be considered as a little magnet.

The key phrase here is "as if." Our argument shows that this spin of $\frac{1}{2}$ cannot be due to the literal rotation of the electron. But this just means that there are other dimensions of reality than those of ordinary space and time. The electron simply has its own internal degrees of freedom. We can, for example, construct a new two-dimensional space, with two axes: one for spin $+\frac{1}{2}$ and one for $-\frac{1}{2}$. The state vector can take positions either on an axis or off. If we wish to combine data concerning spin and data concerning, say, position, in one space, so that one state vector will give all the information, what we do is simply make two copies of our infinite-dimensional linear space, and label one set of spatial axes "spin up" (meaning spin $+\frac{1}{2}$; or, rather, $+\frac{1}{2}\hbar$) and one set "spin down." Another way to put this is that the psi-function has two components, $\Psi_{up}(x)$ and $\Psi_{down}(x)$. The former function (absolute squared) gives the probability of finding the electron at position x with spin "up," while the latter gives the probability to find the electron at position x and also spin "down."

It must be admitted, therefore, that the Maximality Principle, in its original formulation, fails. Even complete knowledge of position amplitudes will not determine the spin amplitude of the electron. Yet the reason for this failure is not so much the principle itself, but the narrowness of its application. By the mistaken assumption that elementary particles must "live" in three-dimensional space of macroscopic experience, we have truncated our linear space by half. Yet even our truncated formalism retains enough information to hint at its own extension.

The above discussion, I must stress one more time, does not follow the historical order of discovery. The hypothesis that the electron has a tiny intrinsic "spin"—distinct from its orbit around the nucleus of an atom—was introduced to explain spectroscopic data. Only later did Pauli discover the appropriate extension of the quantum formalism that could encompass the new discovery. Even from hindsight, though, it is remarkable how the reasoning could have gone the other way—how the formalism contains, latently, further developments. This is why the physicist rightly marvels at the applicability of the Hilbert space formalism.

Appendix B

Nucleon–Pion Scattering

Consider a scattering experiment, described in Sternberg 1994, in which pions collide head-on with nucleons at known momenta and energy; the only variables are the *kind* of incoming particles. We set up a particle detector at some fixed nonzero angle α from the path of particles. Again, the only variables are the kind of outcoming particles; momentum and energy are now fixed. Thus, as above, we can describe nucleons, pions, and deltas, respectively, in terms of two-, three-, and four-dimensional spaces. Then Pythagorean reasoning suggests that a combination of pion and a nucleon is *physically* equivalent to a superposition of a delta and a nucleon, as above. In fact, we can do better. The symmetry-preserving isomorphism associates (as a mathematical theorem)

(a) $$\pi^- \otimes p \leftrightarrow \sqrt{1/3}\ \Delta^0 + \sqrt{2/3}\ n.$$

Thus a π^-/proton pair is physically equivalent to the superposition of a neutral delta and a neutron, the coefficients being as shown. Another equivalence is

(b) $$\Delta^0 \leftrightarrow \sqrt{1/3}\ \pi^- \otimes p \oplus \sqrt{2/3}\ n.$$

There is, therefore, a mathematical possibility that the system could do the following:

$$\pi^- p \to \Delta^0 \to \pi^- p,$$

that is, form a Δ^0 which then decays back into a π^-/proton combination. The probability for this theoretical reversal is calculated by multiplying the coefficient of Δ^0 in (a), by the coefficient of $\pi^-\otimes p[\text{roton}]$ in (b), and then squaring the result:

$$\left(\sqrt{1/3} \times \sqrt{1/3}\right)^2 = 1/9.$$

Suppose we start with a $\pi^+ p$ combination, for which we have

(c) $\qquad\qquad\qquad \pi^+\otimes p \leftrightarrow \Delta^{++}$

(the only way we can have charge conservation in this setup).

Then the probability that a $\pi^+ p$ combination will return to itself, even if it forms a delta particle, is 1, i.e., nine times the probability that the $\pi^- p$ combination will return to itself if it forms a delta particle.

Now all of these calculations have so far no physical significance. We are speaking so far of "incoming" nucleons and pions, i.e., particles which are still far apart. In other words, we have not specified any experiment which realizes this mathematical possibility, though the backwards symmetry condition suggests there should be such an experiment.

So let's allow the particles to interact and then scatter. Recall we are waiting at angle α to see whether particles do scatter. (We are also in the "center-of-mass frame" so that the only momentum variable is the scattering angle itself.) Generally speaking, if particles approach each other head-on, we would expect them either to miss or to rebound along their line of approach. On the other hand, if a nucleon and a pion combine to form a delta which then decays, the debris scatters in any direction. Empirically it is discovered that this happens when the energy of the collision is 180 million electron volts.

Now a remarkable theorem called Schur's Lemma guarantees that the 9 to 1 calculation we have performed remains true even after the pion interacts with the nucleon, even after scattering, so long as this interaction preserves the symmetry condition. Essentially, Schur's Lemma says that, under symmetry, a superposition of two particles remains a superposition. For example, consider the right-hand side of equation (a), which says that a $\pi-p$ pair is a mathematical superposition. At 180 MeV the superposition is actualized, so that we really have the *physical* superposition of a (neutral) delta particle with a neutron. The neutron, if formed, does not scatter, so it has effectively zero chance of getting to our detector. But the delta decays in any direction into a pion/nucleon pair according to the two possibilities implicit in equation (b).

On the other hand, a π^+p pair, at 180 MeV, simply turns into a Δ^{++}, which just decays back into π^+p. The conclusion is, assuming only that the interaction of hadrons has SU(2) symmetry, that at 180 MeV (or whatever energy it takes to form a delta out of a nucleon–pion combination), our detector at scattering angle alpha is nine times more likely to detect the pair π^+p than the pair π^-p.

Appendix C

Nonrelativistic Schroedinger Equation with Spin

Beginning with the equation

$$\left(E^{1/2} - \frac{\sigma \cdot p}{(2m)^{1/2}} \right)\left(E^{1/2} + \frac{\sigma \cdot p}{(2m)^{1/2}} \right) = 0,$$

we make the usual substitutions

$$E \rightarrow i\hbar \frac{\partial}{\partial t} eV$$

$$p_r \rightarrow -i\hbar \frac{\partial}{\partial x_r} - \frac{e}{c} A_r \ (r = 1, 2, 3).$$

In the case where we are not interested in an electromagnetic field that changes over time, we can make for E the simpler substitution

$$E \rightarrow eV.$$

Remultiplying, we recall that the magnetic *field* components B_r are related to the magnetic potential A_r by

$$\mathbf{B} = \nabla \times \mathbf{A},$$

i.e.,

$$B_x = \frac{\partial}{\partial y} A_z - \frac{\partial}{\partial z} A_y$$

$$B_y = \frac{\partial}{\partial z} A_x - \frac{\partial}{\partial x} A_z$$

$$B_z = \frac{\partial}{\partial x} A_y - \frac{\partial}{\partial y} A_x.$$

We note, too, that the Pauli matrices anti-commute

$$\sigma_i\sigma_j = -\sigma_j\sigma_i \,(i \neq j),$$

and that their square is 1:

$$\sigma_i\sigma_i = I = \begin{pmatrix} 1 & 0 \\ 0 & 1 \end{pmatrix} (i = 1, 2, 3).$$

Performing the multiplication, and supplying the psi-function, we derive the following equation for an electron in an electromagnetic potential:

$$\frac{1}{2m} \sum_{r=1}^{3} \left(-\hbar\frac{\partial}{\partial x_r} - \frac{e}{c}A_r \right)^2 \Psi - \frac{e}{m}\frac{1}{c}\frac{\hbar}{2} - \sigma\cdot\mathbf{B}\Psi - (E-eV)\Psi = 0.$$

References

Albert, David Z. 1992. *Quantum Mechanics and Experience*. Cambridge, Mass.: Harvard University Press.

Benacerraf, Paul. 1981. "Frege: The Last Logicist." *Midwest Studies in Philosophy* 6:17–35.

Benacerraf, Paul, and Hilary Putnam, eds. 1984. *Philosophy of Mathematics: Selected Readings*. 2nd edn. Cambridge, England: Cambridge University Press.

Boolos, George. 1987. "The Consistency of Frege's Foundations of Arithmetic." *On Being and Saying: Essays for Richard Cartwright*, ed. Judith Jarvis Thomson, 3–20. Cambridge, Mass.: MIT Press.

Born, M., W. Heisenberg, and P. Jordan. 1925. *Zeitschrift für Physik* 35: 557.

Brague, Rémi. 1990. "*Le Géocentrisme comme humiliation de l'homme.*" *Herméneutique et Ontologie*, 203–33. Paris: Presses Universitaires de France.

Cartwright, Nancy. 1983. *How the Laws of Physics Lie*. Oxford: Clarendon Press.

Chandler, Bruce, and William Magnus. 1982. *The History of Combinatorial Group Theory: A Case Study in the History of Ideas*. New York: Springer-Verlag.

Darrigol, Olivier. 1992. *From c-Numbers to q-Numbers*. Berkeley and Los Angeles: University of California Press.

Detlefsen, Michael. 1992. "Brouwerian Intuitionism." *Proof and Knowledge in Mathematics*, ed. Michael Detlefsen. Oxford and New York: Oxford University Press.

Dirac, P. M. A. 1926. "The Fundamental Equations of Quantum Mechanics." *Proceedings of the Royal Society*, Ser. A, vol. 109:642–53.

———. 1927. "The Quantum Theory of the Emission and Absorption of Radiation." *Proceedings of the Royal Society of London*, Ser. A, vol. 114.

Reprinted in *Selected Papers on Quantum Electrodynamics*, ed. Julian Schwinger. New York: Dover, 1958, 1–23.

———. 1958. *The Principles of Quantum Mechanics*. Oxford: Clarendon Press.

———. "The Relativistic Wave Equation." *European Physics News* 8.10 (Oct. 1977):1–4.

Doncel, Manuel, et. al. 1987. *Symmetries in Physics (1600–1980)*. Barcelona: Servei de Publicacions, Universitat Autonoma de Barcelona.

Drechsler, W. and M. E. Mayer. 1977. *Fibre Bundle Techniques in Gauge Theories*. New York: Springer-Verlag.

Duhem, Pierre. 1962. *The Aim and Structure of Physical Theory*. Trans. from the French by Philip P. Wiener. New York: Anatheum.

Dummett, Michael. 1991a. *Frege: Philosophy of Mathematics*. Cambridge, Mass.: Harvard University Press.

———. 1991b. *Frege and Other Philosophers*. Oxford: Oxford University Press.

———. 1994. "What is Mathematics About?" In *Mathematics and Mind*, ed. Alexander George, 11–26. Oxford: Oxford University Press.

Dyson, F. J. 1969. "Mathematics in the Physical Sciences." In *The Mathematical Sciences*, ed. Committee on Support of Research in the Mathematical Sciences (COSRIMS) of the National Research Council, 97–115. Cambridge, Mass.: MIT Press.

Einstein, Albert. 1974. *The Meaning of Relativity*. 4th paperback edn. Princeton: Princeton University Press.

Feyerabend, Paul. 1978. *Against Method*. London: Verso.

Feynman, Richard. 1967. *The Character of Physical Law*. Cambridge, Mass.: MIT Press.

Feynman, Richard P., and Steven Weinberg. 1987. *Elementary Particles and the Laws of Physics*. Cambridge, England: Cambridge University Press.

Field, Hartry. 1980. *Science without Numbers: A Defense of Nominalism*. Princeton: Princeton University Press.

———. 1989. *Realism, Mathematics, and Modality*. Oxford: Blackwell.

Freischmann, M.D., J. Tildesley, and R.C. Ball, eds. 1990. *Fractals in the Natural Sciences*. Princeton: Princeton University Press.

Frege, Gottlob. 1959. *The Foundations of Arithmetic*. Trans. by J.L. Austin. Oxford: Blackwell, republished by Northwestern University Press.

French, A. P., and Edwin F. Taylor. 1978. *An Introduction to Quantum Physics*. New York: W. W. Norton Inc.

Galindo, A., and C. Sanchez Del Rio. 1961. "Intrinsic Magnetic Moment as a Nonrelativistic Phenomenon." *American Journal of Physics* 29:582–89.

Gell-Mann, Murray. 1987. "Particle Theory from S-Matrix to Quarks."

In *Symmetries in Physics (1600–1980)*, ed. Manuel Doncel et al., 473–97. Barcelona: Servei de Publicacions, Universitat Autonoma de Barcelona.

Gell-Mann, Murray, and Yuval Ne'eman, eds. 1964. *The Eightfold Way*. New York: W. A. Benjamin.

Goldstein, Herbert. 1950. *Classical Mechanics*. Reading: Addison-Wesley.

Goodman, Nelson. 1983. *Fact, Fiction, and Forecast*. 4th edn. Cambridge, Mass.: Harvard University Press.

Graves, John C. 1971. *The Conceptual Foundations of Contemporary Relativity Theory*. Cambridge, Mass.: MIT Press.

Guillemin, Victor, and Shlomo Sternberg. 1990a. *Symplectic Techniques in Physics*. Reprinted with corrections from 1984 edition. Cambridge: Cambridge University Press.

———. 1990b. *Variations on a Theme by Kepler*. Providence, R.I.: American Mathematical Society Colloquium Publications, vol. 42.

Hardy, G. H. 1967. *A Mathematician's Apology*. Cambridge, England: Cambridge University Press.

Harman, Gilbert H. 1989. "The Inference to the Best Explanation." In *Readings in the Philosophy of Science*, ed. Baruch A. Brody and Richard E. Grandy, 2nd edn., 323–28. Englewood Cliffs, N.J.: Prentice Hall.

Harper, William. 1990. "Newton's Deductions from Phenomena." *PSA* 2:183–96.

Hawking, Steven. 1988. *A Brief History of Time*. Toronto: Bantam.

Heisenberg, Werner. 1925. *Zeitschrift für Physik* 33:879.

Hellman, Geoffrey. 1989. *Mathematics without Numbers*. Oxford: Clarendon Press.

Henry, Richard C. 1990. "Quantum Mechanics Made Transparent." *American Journal of Physics* 58.11 (Nov.):1087–100.

Hersey George. 1976. *Pythagorean Palaces*. Ithaca, N.Y.: Cornell University Press.

Hunt, Bruce J. 1991. Ithaca, N.Y.: Cornell University Press.

Irvine, A.D., ed. 1990. *Physicalism in Mathematics*. Dordrecht: Kluwer Academic Publishers.

Kac, Mark, and Stanislaw Ulam. 1971. *Mathematics and Logic*. Harmondsworth, England: Penguin Books.

Kant, Immanuel. 1950. *Prolegomena to Any Future Metaphysics*. Indianapolis: Liberal Arts Press.

Kitcher, Philip. 1983. *The Nature of Mathematical Knowledge*. New York: Oxford University Press.

Kripke, Saul. 1980. *Naming and Necessity*. Cambridge, Mass.: Harvard University Press.

———. *Wittgenstein on Rules and Private Language*. Cambridge, Mass.: Harvard University Press, 1982.

Landau, L.D., and E.M. Lifshitz. 1965. *Quantum Mechanics: Non-relativistic Theory*. 2nd (revised) edn. Trans. from the Russian by J.B. Sykes and J.S. Bell. Oxford: Pergamon Press.

Levi, Isaac. 1980. *The Enterprise Of Knowledge*. Cambridge, Mass.: MIT Press.

Levy-Leblonde, J.-M. 1979. "The Importance of Being (A) Constant." *Problems in the Foundations of Physics* 72:238–63.

Lovejoy, Arthur O. 1964. *The Great Chain of Being*. Cambridge, Mass.: Harvard University Press.

Mac Lane, Saunders. 1986. *Mathematics: Form and Function*. New York: Springer-Verlag.

Mandelbrot, B.B. 1990. "Fractal Geometry: What is it and What does it Do?" In *Fractals in the Natural Sciences*, ed. M. Fleischmann, D.J. Tildesley, and R.C. Ball, 3–16. Princeton: Princeton University Press.

Mandelbrot, Benoit B., and Carl J.G. Evertsz. 1990. "The Potential Distributions around Growing Fractal Clusters." *Nature* 348.6297 (8 Nov.):143–5.

Manin, Yu.I. 1981. *Mathematics and Physics*. Boston, Basel, and Stuttgart: Birkhauser.

Maxwell, James Clerk. 1954. *A Treatise on Electricity and Magnetism*. 3rd edn. New York: Dover.

Messiah, Albert. 1962. *Quantum Mechanics*. Amsterdam: North-Holland.

Mills, Robert. 1989. "Gauge Fields." *American Journal of Physics* 57.6 (June):493–507.

Nagel, Ernest. 1979. "Impossible Numbers." In *Teleology Revisited*, 194. New York: Columbia University Press,

Ne'eman, Yuval. "Hadron Symmetry, Classification, And Compositeness." In *Symmetries In Physics (1600–1980)*, ed. Manuel Doncel, et. al., 501–24. Barcelona: Servei de Publicacions, Universitat Autonoma de Barcelona.

Newton, Isaac. 1934. *Mathematical Principles of Natural Philosophy*. Berkeley: University of California Press.

O'Meara, Dominic J. 1989. *Pythagoras Revived: Mathematics and Philosophy in Late Antiquity*. Appendix: The Excerpts from Iamblichus' *On Pythagoreanism* V–VII in Psellus: Text, Translation, and Notes. Oxford: Clarendon Press.

Pais, Abraham. 1986. *Inward Bound*. Oxford: Oxford University Press.

Parsons, Charles. 1964. "Frege's Theory of Number." In *Philosophy in America*, ed. Max Black, 180–203. Ithaca, N.Y.: Cornell University Press.

Pauli, Wolfgang. 1926. "Uber Das Wasserstoffspektrum vom Standpunkt der Neuen Quantenmechanik." *Zeitschrfit für Physik* 36:336–63. Translation in Van Der Waerden 1967.

Peirce, C.S. 1958. *Collected Papers Of Charles Sanders Peirce*, ed. Arthur W. Burks. Vol. 7. Cambridge, Mass.: Harvard University Press.

Penrose, Roger. 1978. "The Geometry of the Universe." In *Mathematics Today: Twelve Informal Essays*, ed. Lynn Arthur Steen, 83–125. New York: Springer-Verlag.

————. 1989. *The Emperor's New Mind*. New York: Oxford University Press.

Pour-El, Marian B., and J. Ian Richards. 1989. *Computability in Analysis and Physics*. Berlin: Springer-Verlag.

Primas, Hans. 1983. *Chemistry, Quantum Mechanics and Reductionism*. Berlin: Springer-Verglag.

Quine, W. V. 1960. *Word and Object*. Cambridge, Mass.: MIT Press.

Reed, Michael, and Barry Simon. 1975. *Methods Of Modern Mathematical Physics*. Vol. II. New York: Academic Press.

Rescher, Nicholas. 1984. *The Riddle of Existence*. Lanham: University Press of America.

Ruelle, David. 1991. *Chance and Chaos*. Princeton: Princeton University Press.

Schroedinger, Erwin. 1978. *Collected Papers on Wave Mechanics*. New York: Chelsea.

Schwinger, Julian, ed. 1958. *Selected Papers on Quantum Electrodynamics*. New York: Dover.

Shapiro, Stewart. 1984. "Mathematics and Reality." *Philosophy of Science* 50:523–48.

Shenker, Orly. 1994. "Fractal Geometry is Not the Geometry of Nature." *Studies in the History and Philosophy of Science* 25:967–81.

Siegel, Daniel M. 1991. *Innovation in Maxwell's Electromagnetic Theory*. Cambridge, England: Cambridge University Press.

Steiner, Mark. 1986. "Events and Causality." *Journal of Philosophy* 83:249–64.

————. 1989. "The Application of Mathematics to Natural Science." *Journal of Philosophy* 86:449–80.

————. 1995. "The Applicabilities of Mathematics." *Philosophia Mathematica* 3:129–56.

Sternberg, Shlomo. 1994. *Group Theory and Physics*. Cambridge, England: Cambridge University Press.

Van Der Waerden, B.L., ed. 1967. *Sources of Quantum Mechanics*. Amsterdam: North-Holland.

Van Fraassen, Bas C. 1980. *The Scientific Image*. Oxford: Oxford University Press.

von Neumann, John. 1956. "The Mathematician." In *The World of Mathematics*, ed. James R. Newman, 2053–2063. New York: Simon and Schuster.

Weinberg, Steven. 1986. "Lecture on the Applicability of Mathematics." *Notices of the American Mathematical Society* 33.5 (Oct.).

Wells, David. 1988. "Which is the Most Beautiful?" *Mathematical Intelligencer* 10:30–1.

Wessels, Linda. 1977. "Schroedinger's Route to Wave Mechanics." *Studies in the History and Philosophy of Science* 10:311–40.

Weyl, Hermann. 1950. *The Theory of Groups and Quantum Mechanics.* Trans. from the 2nd revised German edn. by H. P. Robertson. New York: Dover.

———. 1952. *Symmetry.* Princeton: Princeton University Press.

Wightman, A.S. 1969. "Analytic Functions and Elementary Particles." In *The Mathematical Sciences*, ed. The Committee on Support of Research in the Mathematical Sciences (COSRIMS) of the National Research Council, 116–27. Cambridge, Mass.: MIT Press.

Wigner, Eugene. 1967. "The Unreasonable Effectiveness of Mathematics in the Natural Sciences." In *Symmetries and Reflections*, 222–37. Bloomington: Indiana University Press.

Wilczek, Frank. 1991. "Anyons." *Scientific American* 264.5 (May):58–65.

Wilson, Mark. 1992. "Frege: The Royal Road from Geometry." *Nous* 26:149–80.

Wittgenstein, Ludwig. 1968. *Philosophical Investigations.* 3rd edn. Trans. G.E.M. Anscombe. Oxford: Basil Blackwell; New York: Macmillan.

———. 1978. *Remarks on the Foundations of Mathematics.* Ed. G. H. Von Wright, R. Rhees, and G.E.M. Anscombe, revised edn. Cambridge, Mass.: MIT Press.

Yamanouchi, T. 1970. "Quantum Mechanics." In *Mathematics Applied to Physics*, ed. E. Roubine, 562–601. Berlin: Springer-Verlag.

Yang, C.N. 1952. "The Spontaneous Magnetization of a Two-dimensional Ising Model." *Physical Review* 85.5 (Mar. 1). Reprinted in Yang 1983, 142–50.

———. 1977. "Magnetic Monopoles, Fiber Bundles, and Gauge Fields." *Annals of New York Academy of Sciences* 294 (Nov. 8):86–97.

———. 1983. *Selected Papers 1945–1980 with Commentary.* New York: W.H. Freeman and Company.

Yang, C.N., and R.L. Mills. 1954. "Conservation of Isotopic Spin and Isotopic Gauge Invariance." *Physical Review* 96:191–5.

Zahar, Elie. 1989. *Einstein's Revolution: A Study in Heuristic.* La Salle, Ill.: Open Court.

Zee, A. 1990. "The Effectiveness of Mathematics in Fundamental Physics."

In *Mathematics and Science*, ed. Ronald E. Mickens, 307–23 Singapore: World Scientific.

Zhang, D.Z. 1993. "C.N. Yang and Contemporary Mathematics." *The Mathematical Intelligencer* 15.4 (Fall):13–21.

Index